Teacher Learning and Informal Science Education

Studies in Criticality

Shirley R. Steinberg
Series Editor

Vol. 549

Jennifer D. Adams

Teacher Learning and Informal Science Education

Expansivising Affordances for Diverse Science Learners

PETER LANG
New York - Berlin - Bruxelles - Chennai - Lausanne - Oxford

Library of Congress Cataloging-in-Publication Data

Names: Adams, Jennifer D., author.
Title: Teacher learning and informal science education: expansivising affordances for diverse science learners / Jennifer D. Adams.
Description: New York: Peter Lang, [2024] | Series: Counterpoints, 1058–1634; vol. 549 | Includes bibliographical references.
Identifiers: LCCN 2023054671 (print) | LCCN 2023054672 (ebook) | ISBN 9781636672830 (paperback) | ISBN 9781636672847 (hardback) | ISBN 9781636672816 (pdf) | ISBN 9781636672823 (epub)
Subjects: LCSH: Science–Study and teaching–Case studies. | Minorities–Education--Case studies. | Science teachers–Training of. | Minority teachers–Training of. | Communication in science.
Classification: LCC Q181. A2895 2024 (print) | LCC Q181 (ebook) | DDC 507.1–dc23/eng/20240201
LC record available at https://lccn.loc.gov/2023054671
LC ebook record available at https://lccn.loc.gov/2023054672

Bibliographic information published by the Deutsche Nationalbibliothek.
The German National Library lists this publication in the German National Bibliography; detailed bibliographic data is available on the Internet at http://dnb.d-nb.de.

Cover design by Peter Lang Group AG

ISSN 1058-1634 (print)
ISBN 9781636672830 (paperback)
ISBN 9781636672847 (hardback)
ISBN 9781636672816 (ebook)
ISBN 9781636672823 (epub)
DOI 10.3726/b21130

© 2024 Peter Lang Group AG, Lausanne
Published by Peter Lang Publishing Inc., New York, USA
info@peterlang.com - www.peterlang.com

All rights reserved.
All parts of this publication are protected by copyright.
Any utilization outside the strict limits of the copyright law, without the permission of the publisher, is forbidden and liable to prosecution.
This applies in particular to reproductions, translations, microfilming, and storage and processing in electronic retrieval systems.

This publication has been peer reviewed.

To Prince (Flash) and Carmel (Peggy) Adams.
With love and gratitude.

CONTENTS

	Figures	ix
	Abbreviations	xi
Chapter 1	Expansivising Spaces and Affordances for Science Learning	1
Chapter 2	An Artifact of Settler Colonialism	33
Chapter 3	Teacher Agency and Identity: Creating Affordances in the Expansivising Space	65
Chapter 4	Modifying Resources: Adapting Museum Display and Investigation for the Classroom	93
Chapter 5	Expanded Agency Through Modifying and Creating Affordances	109
Chapter 6	Critically Expansivising Practices: Teachers as Bricoleurs	127
Chapter 7	Challenging "Scientific" Research Paradigms Through Sociocultural Lenses and Dialogic Methodologies	155
Chapter 8	Youth Practices as Expansivising Resources: DWL on WhatsApp	173

FIGURES

Figure 1.1. Expansivising space between museums and schools where schema and resources from each come together to form new learning cultures. Examples of the schema and resources found in each context. The dotted lines indicate that the schema and resources are neither fixed nor bounded to a particular context—they move across membranes of contexts.   11

Figure 2.1. The Theodore Roosevelt statue displays racist, eugenicist notions of racial hierarchy in bronze. Roosevelt represents the pinnacle of Man with both African and Indigenous humans dysselected from that category.   37

Figure 2.2. Children studying nature in the Museum. © The American Museum of Natural History   57

Figure 3.1. In the expansivising space new learning cultures are formed and identities confirmed/evolved as resources and schema from different learning cultures, in this case ISE and the formal classroom, come into relation. Human agency and identity shape the affordances and the ways that the affordances become enactments (immediate actions) and eventually repertoires of practices (patterns of enactments).   69

Figure 3.2. Applying the UMC model (Lee et al., 2014) to teacher learning. Each enactment through the UMC cycle contributes to establishing repertoires of practice, expanding teaching agency and corresponding identity.   75

ABBREVIATIONS

AMNH   American Museum of Natural History
CAB   Critical Agentic Bricoleur
CBMS   Central Brooklyn Middle School
DTP   Decolonize This Place
ILETES   Informal Learning Environments and Teacher Education for STEM
ISE   Informal Science Education
ISI   Informal Science Institution
MLC   Museum Learning Collaborative
NCLB   No Child Left Behind
NYCMS New York City Museum School

· 1 ·

EXPANSIVISING SPACES AND AFFORDANCES FOR SCIENCE LEARNING

Informal science institutions (ISIs) are spaces designed for science learning that center public engagement with science. These institutions include not only museums, but also science centers, nature centers, botanical gardens, zoos, and aquariums—all contain objects, artifacts, activities, and stories related to science. The public-facing spaces, the halls and exhibits, are designed to display and communicate the scientific phenomena. Whether static or interactive, these displays are curated to tell a story and/or convey a specific meaning around science. For example, many zoos focus on animal conservation so, in addition to facts about animals on display, there may also be information about their conservation status or stories about scientists and community members who work closely with the animals. Many science-rich cultural institutions also have inaccessible areas dedicated to scientific knowledge production. These include laboratories, the collections of objects, artifacts, and specimens, and education departments that design learning resources and facilitate learning. Worldwide, ISIs play an important role in shaping public knowledge and ideas about science.

Because of the varying contexts for learning, even within one institution, research on learning in informal science institutions has helped to inform broader knowledge about how people learn science. Studies on informal science learning range from examining visitor interests and motivations to analyzing family dialogues in exhibits and examining school group interactions. These have allowed the field to gain deeper understandings about how people learn and helped broaden pedagogical approaches to science teaching and learning both in ISIs and in formal classrooms.

As we become more aware of the ways a culture of whiteness has structured museums, informal science institutions, and learning in general, there is a corresponding advance in research into peoples' larger ecologies of science learning with an emphasis on non-Western, decolonial ways of knowing and doing science. Such explorations naturally lend themselves to the less structured, nonlinear and often interdisciplinary spaces of ISIs.

In this chapter, I briefly describe the theoretical and historical foundations of informal science education (ISE), including my own navigations of theories of learning, to better understand teacher learning in museum/ISI settings. As a museum educator at the American Museum of Natural History, I was able to observe the ways that partnerships shaped science teaching and learning within the museum and in partnering schools. I describe these experiences and propose a theoretical perspective of "expansivising spaces" where new learning cultures and affordances for science learning are created. I then discuss how teachers learn in these expansivising spaces as foundational to describing and expanding on teacher learning throughout this book.

My Entry into Museum Education

I started my tenure as a museum educator after teaching high school biology and Earth science for almost a decade. My role at the American Museum of Natural History (AMNH) was to design and facilitate professional learning for teachers. Based on my teaching experiences, I brought two things to the planning table. The first was that aspects of professional learning had to be immediately applicable—this meant the teachers had to learn a strategy, activity, content connection, etc., that they could immediately use in the classroom with little modification. As a teacher, I was always searching for activities that would make science more engaging: hands-on activities that also allowed students to dialogue with each other while learning and applying science concepts and processes. Having immediately applicable activities available for teachers in association with professional development reduced the burden of them searching for activities and allowed more room for the discussion of integrating the activity into the continuum of classroom practices. The second thing centered in my interactions was that teacher learning also had to consider the teachers' longer-term professional identities. What are the strategies and stances introduced in teacher learning that contribute to teachers' development of repertoires of practices? What is foundational to the ongoing development of engaging and equitable science teaching and learning practices? These are important considerations because teachers develop practices that resonate with their professional identities: this includes visions of themselves and how they want others to view them as teachers. Because of the visual and multi-modal cultures of ISIs, we emphasized two practices: observation and inquiry. We positioned observation is a way of learning and knowing, and observations could form the basis of curiosity—wonderings and questions learners could then further explore.

Prior to working at the museum, one of the schools I worked at was a considerable distance from Manhattan (where the museum is located). The school was in my Brooklyn community, one that was predominantly Caribbean (Afro, Indo and mixed) with sprinklings of Latinx and African Americans, recent Polish and Russian immigrants, as well as a few Italian, Irish and Jewish holdouts from when the community was predominantly white. I was able to bring my classes to the museum a couple of times, but it was usually a logistical challenge in ensuring the timing of school buses and/or negotiating various means of public transportation to get my 30–50 students to and from the museum. However, I viewed these museum visits as important in providing a different experience for the students from what they usually experienced in the classroom. The museum would be a way to enrich what they learned from recipe labs, worksheets, and textbook assignments. At the time, as a new teacher with limited pedagogical training, I ended up repeating what I did in the classroom at the museum—I gave the students a worksheet to complete when we visited the Hall of Human Biology. I had them watch the films and look at the exhibits for "answers" to questions and prompts on the worksheets. Reflecting on that activity now, it was boring AF, but my students were engaged nonetheless and completed the worksheet. Visiting the museum was a different experience: it got them out of the building and borough and allowed them to engage with me, each other, and the exhibits in ways that were not possible in the classroom space. That seemed like enough motivation to get them to do the work, much more than they ever did with any similar worksheet activities in the classroom.

After spending time in one (museum) hall on an activity, I usually gave students the opportunity roam any hall in the museum they desired (in pairs or small groups). I would head to the Hall of Ocean Life because of the iconic and easily recognizable blue whale. This would also serve as the meeting spot. I was always surprised by how many of the students followed me to this hall. My guess is that even as "fly[1]" high schoolers, they felt more comfortable with me in this unfamiliar place (some of them never visited AMNH before, and for others, they hadn't visited since elementary school). For some, they wanted to learn more and by staying with me, I would help to mediate some of their questions and interactions with the exhibits.

I always enjoyed these trips and the opportunity to stand back and watch my students get excited upon encountering new objects and facts. Although my worksheet was didactic, the practice of observation was still key—the students had to look carefully or observe to find the answers. Watching the students interact in the halls, I felt like how my father must have felt when he took us to the museum when we were children and saw our eyes widen at the sight of the same blue whale and dinosaurs on the fourth floor. One of the ways my father showed his care for us was by taking us to the museum, and that act of care repeated itself when I ventured into the museum with my students. Returning to this museum as an educator completed the cycle of care. I now had the opportunity to support teachers in using museum resources to deepen and enrich science teaching and learning for their students.

My Initial Research on Learning in the Museum

Up until this point, my research experiences in postsecondary education and as a biology teacher were lab-based experiments that required control and experimental models. Similarly, my science education courses emphasized teaching and learning science content without delving too much into theories of learning. Since I entered education via an alternative route, I have not yet been exposed to the education canon and theories of learning. I had course experiences in psychology, which influenced how I viewed the process of learning as a brain-based cognitive process. I only had emergent academic understandings about how culture and identity shape learning, even though my experiences as a teacher allowed me to realize that making science learning relevant to students' lives was the best way to keep them engaged.

One of my early research projects on museum learning was in the context of an educational psychology course. I focused on children's literacy and museums with the question of how a child's brain would process the museum learning experience. At the same time, the *No Child Left Behind* mandates of the Bush era were taking hold, and schools were placing greater emphasis on literacy and math—the standardized testing subjects. Class visits to the museum plummeted because of these mandates, and we were concerned with "justifying" the benefit of museum visits to subject areas beyond science. Connecting museum visits to literacy would provide the needed evidence that museum visits could contribute to both English and science literacy.

This was an exploratory study, and since it employed the usual educational practices, I did not need to study ethics at the time. I leveraged convenient participants consisting of my two young nieces and a colleague's daughter, all in the third or fourth grade. I designed a study where I took them to a specific display about the deep ocean ecosystem and engaged them with learning about the display through questioning and observations in the same way I would when modeling how to use the display for teachers. Approximately a week later, they read a related text titled *Creeps of the Deep*, while I used the think-aloud protocol (Kucan & Beck, 1996) to determine whether they visualized images from the museum while they were reading the text. Would they show evidence of incorporating background knowledge gained from viewing the display when they described their understanding of the text?

I leveraged Sadoski et al.'s (1991) theory of dual coding, which describes cognition as the activity of two separate mental "subsystems": one for the representation and processing of nonverbal objects and events (imagens) and the other for language (logogens). Because memory is processed and stored in two mental subsystems, my interpretation was that when people learn in places that allow for multisensory experiences, such as museums, they could process and store information better since each working memory system would not be overloaded. Yet, as the think-aloud unfolded, there were very few direct references to the museum display. Rather, the participants' memories of their own interactions with the ocean came to mind. For example, in response to the sentence, "Below the sunlight surface layer, the sea gets dim, dimmer and then dark," my niece responded, "It makes me think that the ocean is getting dark and it's like in the deep sea when we went to the dock when we got the fish." Here, her experience with a dark ocean was going fishing at night, and her understanding of the deep ocean is that fish "live" there. My colleague's daughter referred to piranhas in relation to the deep ocean, recounting a visit she had to the zoo with her parents. The humpback anglerfish of the deep sea has a similar shape and teeth to the piranha, hence the possible connection, but mentioning both the experience and the people in relation to a scientific phenomenon was important.

This small think-aloud project allowed me to make some salient inferences about museum learning and science learning more broadly: learners' experiences will influence how they interact with and understand scientific phenomena. For the three young learners, the experiences with their families in relation to the topic sparked their imagination in relation to the text. It would be challenging to parse their understanding into the deposit and retrieval of imagens and logogens, as this dual processing system implies. These theories produce a "banking" design for learning where educators "make deposits which the students patiently receive, memorize, and repeat" (Freire, 1970, p. 72).

Textual and visual prompts are not separate systems but entangled in the complexity of human knowing, living, and understanding, in addition to ongoing knowledge-building and sharing. This research experience confirmed my growing understanding of learning as complex and intertwined psychological, sociocultural, and neurological processes.

Many of these cognitive theories emerged from experimental settings (in a laboratory setting with control and experimental or treatment groups), where there is presumed scientific neutrality and objectivity, often resulting in "static definitions" of learning that do not consider how "unique social contexts and institutional affordances of museums might transform meanings of learning" (Leinhardt and Crowley, 1998, p. 3). With the underlying assumption that all brains work the same, it produced ableist learning approaches where brains that do not seem to process things as they were described in laboratory experiments are considered deviant, abnormal, or even less than human. It is also important to consider that many educational models (standards articulation, curriculum development, assessment, tracking, etc.) are based on the theories that emerged from these sorts of experiments. However, as I spent more time in the museum, especially in my role as a teacher educator, I sought to describe learning with the theories that resonated with my experiences, i.e., learning is a complex interplay of various factors not limited to brain-based neutrality.

Explanatory Dialogues and Museum Teaching

As museum educators, we collaboratively planned and facilitated professional development programs and spent time watching each other facilitate learning with teachers in the halls. It was a time to observe and learn from each other both about the exhibit and to experience different facilitation styles. We emphasized using questioning strategies to engage learners with a diorama and display, and as such, I became curious about the ways that professional identity intersected with this activity. Particularly, I was interested in the ways that people with different professional orientations facilitated learning interactions in the museum. With my colleagues having similar curiosities, we engaged in our own inquiry to see if there were differences in how an educator and a scientist would mediate the same exhibit. Using the mangrove display in the Milstein Hall of Ocean Life, I prompted two of my colleagues, a museum educator, and an ichthyologist, to facilitate interaction with the display as they would with a group of teachers. At the time, I was also becoming more familiar with existing museum-based research on learning that emphasized conversations, so I framed my analysis around the idea of instructional explanations—the ways that teachers communicate to support the learning of others ("Instructional explanations are designed to explain concepts, procedures, events, ideas, and classes of problems in order to help students understand, learn, and use information in flexible ways"; Leinhardt, 1997, p. 223). Thus, I looked for the ways that they each used questions to teach and communicate about science in relation to the museum objects.

The museum educator used a narrative approach that allowed the viewer to be a part of a story as it might unfold in the display. She asked questions that prompted teachers to think about what was happening, what specimen was on display, and the relationships and interactions between the specimen and objects on display. Teachers should use the evidence that they viewed in the display, along with their own prior knowledge, to create a narrative about the organisms found in a mangrove. The ichthyologist prompted exploration with the display by using questions that asked teachers to describe the characteristics of the individual objects and how they were adapted to the mangrove. As a fish systematist, the objects represented scientific facts and content; she emphasized the need to know the components to understand the entire system. Meanwhile, with a background in elementary education, the museum educator used storytelling to animate the display. They both used questions that required careful observation and making inferences based on those observations. However, their professional identities became evident in what the questions prompted the teachers to observe.

Although theories of learning (i.e., from psychology) have influenced how people think about learning in museums, the museum education field has endeavored to describe how learning unfolds during museum interactions. These studies, often qualitative, have allowed us to extend learning theories beyond individual cognition toward centralizing sociocultural interactions. Museum-based research on learning often documents learning as it happens through observations, discourse analysis, and interviews/dialogues with visitors and participants. Pioneering this work was the Museum Learning Collaborative (MLC) out of the University of Pittsburgh, where they addressed the need for research to better understand learning in museums. Mobilized around the question, "How does conversation as a socially mediating activity act as both a process and an outcome of museum learning experiences?" (Leinhardt & Crowley, 1998, p. 2), the MLC sought to address the problem of what they viewed as the lack of coherence of existing learning theories with the dynamic, diverse, and highly social nature of learning in museums. The MLC defined learning as conversational elaboration since it is a "naturally occurring and meaningful process and product of the museum experience" (p. 5). This approach recognizes that identity, explanatory engagement, and context affect how conversations build and expand in museums; in other words, learning in museums is a social activity.

Further, realizing that objects are central to the museum experience, in the volume *Perspectives on Object-Centered Learning in Museums* (Paris, 2002), various researchers explored how social interactions, prior knowledge, and experiences influenced the kinds of learning engagements and interpretations people have with and of objects. In the foreword, John Falk (2002) described objects as representing "a vast continuum of abstract ideas and inter-related realities" (p. x). Recognizing that many objects also represent objects that exist outside of the museum, learning with objects must also consider the diverse identities, experiences, and worldviews that people bring to their learning interactions with objects. An object-based epistemology considers the transactions/relations between people and objects as ways of meaning-making through connections, curiosity, and affect (Paris, 2002). In museums, educators often have the role of mediating the learning interactions between visitors and objects on display. In the case of teacher learning, this mediation is also a way to model the different ways teachers can engage their students in object-based learning. Hence, it became interesting for me to understand how different people with different professional identities—educators and scientists—approached the objects on display.

ISE: Theoretically Defining the Field

Informal science education (ISE) is loosely defined as the learning that happens outside of school. This includes museums, "living" museums, such as zoos, aquaria, and botanical gardens, and place-based education and environmental learning contexts, such as parks, nature centers, and other environmentally related venues. The field of ISE emerged from the desire to understand in what ways and to what extent learning happens in museums and science centers. The National Association of Research in Science Teaching established an Ad Hoc Informal Science Education committee in 1999 to codify the organization's positioning regarding out-of-school learning (Dierking, Falk,

Rennie, Anderson & Ellenbogen, 2003). The committee identified several salient characteristics of learning: (a) that it is strongly influenced by prior knowledge, experience, interest, and motivations, (b) it is a cumulative process that occurs over time, and (c) beyond cognition, there are also emotional, material, and bodily experiences that influence learning. These characteristics have expanded how science, technology, engineering, and mathematics (STEM) education researchers think about learning in formal and informal settings. However, the discussion here will continue to emphasize ISE.

One of the seminal theories of informal science learning was forwarded by Lynn Dierking and John Falk (2002), which described ISE as free-choice learning—nonsequential, self-paced and voluntary and accounts for the social nature of learning—the interaction of the individual with his or her sociocultural and physical environments. To highlight the different dimensions of museum learning, they advanced the Contextual Model of Learning (Falk & Dierking, 2004) and highlighted four realms that interact when one is learning in an informal setting—the physical, social, and personal contexts and time across a lifespan. This view of learning recognizes the situatedness of learning in a museum, which includes different interrelated events—conversations, interactions, physical space, and individual processing—that occur in the process of learning or meaning-making in a museum. Specific to learning with objects, theories that describe the distributive nature of learning have been useful in ISE. For example, framing learning from objects within the framework of distributed activity, Rowe (2002) situates objects, as found in museums, within the network of human cognitive activity, citing that "individuals solving problems or interacting with objects alone retain traces of those social origins" (p. 21). This allowed museum learning to move from a didactic to more interactive pedagogies in relation to learning in the halls and with objects.

With increasing interests in ecological approaches to learning and awareness of the intersections of learning across contexts, cultures and lifespans, ISE expanded to encompass peoples' lived experiences with science in their homes, communities, and everyday interactions. This latter expansion allows informal science researchers and educators to think more deeply about ontological aspects of science and science learning, which is a critical turn towards expanding the meanings of science and science education, including identifying and describing evidence of learning. More importantly, this has afforded the creation of spaces for diverse learners to meaningfully participate in and contribute to a broad spectrum of scientific endeavors. Along with this turn, segments of the informal science education and learning science fields are critically interrogating the underlying ideologies of science with questions about the knowledge centered and valued and those that have been systematically excluded.

Describing ISE also begs the question of place or pedagogy. As a field, ISE is the collective of educators, researchers and other practitioners who work to advance science teaching and learning outside of formal schooling. ISE, as a pedagogy, describes the practices of science teaching and learning that happen in informal science learning contexts. Some of the common practices of informal science learning include guided inquiry and inquiry-based learning, object-based learning, and investigation-oriented approaches that strive to mirror the practices and approaches of scientists. Other common practices include structured learning activities and those that are more unique to informal settings, like student choice, collaborative engagements, organic combining of social activities with learning and the physical space and artifacts available to educators and learners. Thus, as a pedagogy, informal science tends towards inquiry-based learning (Adams & McCullough, 2021) and centering learner agency in interests and approaches. Defining informal science learning as a place becomes more complicated as the notion of informal science learning can be enacted in any place, whether it is a school building, an informal science institution, or a home. Although the association with place, such as museums and science centers, has been long established, the expansion of places recognized as places of science learning has been quite broad. However, engaging with the question of place/pedagogy allows us to think about the broader meanings and orientations of science learning. At once, informal science teaching and learning is both place and pedagogy. While there are pedagogies that define science learning in informal science contexts, such as object- and inquiry-based learning, these are not unique to informal science contexts. It is unique that learners in informal science contexts have expanded opportunities to leverage their own personal and cultural affordances to make meaning with the objects and phenomena they encounter.

Shifting Theories Shape Learning

With shifts in views about learning in museums and more sociocultural theories about learning theories more broadly, in the latter 20th century, the museum began to view partnerships with schools to expand student and teacher learning opportunities. Partnerships afford expansivising interactions of learning cultures and resources between the ISIs and schools. The museum's artifacts and scientific and human resources integrate with curricular resources, learning standards, and extended contact with teachers and students to allow for new spaces and opportunities for learning to emerge (Gupta & Adams, 2012). Expansivising refers to the limitless learning possibilities when different resources are put in relation. This provides options for many different configurations of learning and participation in knowledge production for diverse learners. Diverse learners are those who have been marginalized from meaningful and relevant learning opportunities because the cultural, community, linguistic, embodied, and neurological resources and knowledge that they carry to school/formal education-based learning experiences have neither been affirmed nor valued (Figure 1.1).

Teacher education in the museum moved towards a more expansivising approach with Dr. Maritza MacDonald, who served as the Director of Professional Development from 1997 to 2017. During her tenure, she emphasized teacher learning and concurrently developed a range of opportunities and engagements. Partnerships

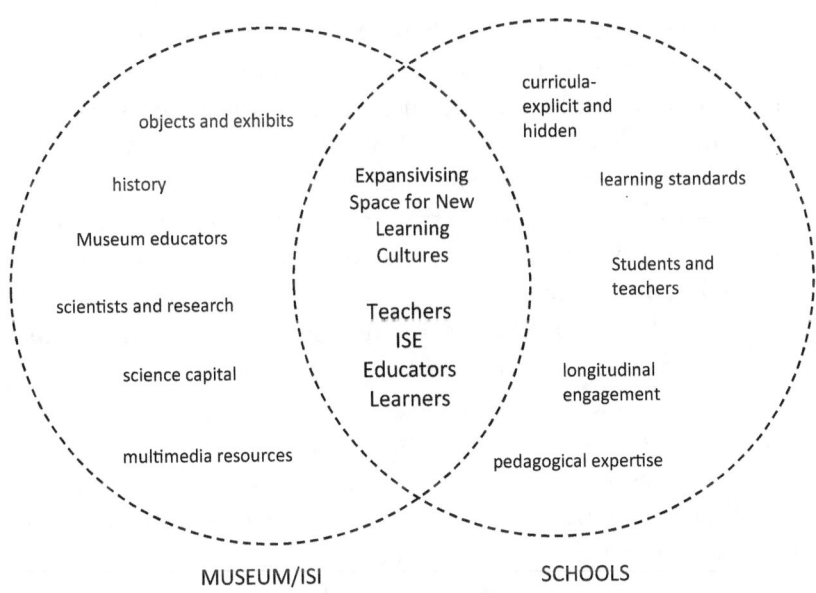

Figure 1.1. Expansivising space between museums and schools where schema and resources from each come together to form new learning cultures. Examples of the schema and resources found in each context. The dotted lines indicate that the schema and resources are neither fixed nor bounded to a particular context—they move across membranes of contexts.

with schools afforded Dr. MacDonald the opportunity to shape museum-based teacher learning around the needs of schools and classrooms. "[Partnerships] give us, as a museum, the opportunity to really understand how a museum can help schools. We deal with the standards and assessments that are guiding schools, so we are more connected to what schools need and do. It teaches us how we prepare ourselves to really help schools over the long haul. We are looking for evidence that we can actually do things for schools."

When I worked at AMNH, "museum as a resource" was a phrase Dr. MacDonald often used when describing the relationship between museum and school science learning. Similarly, the co-founder of one of the schools, Sonnett Takahisa, in partnership with AMNH, noted that "museums are repositories of both the natural world and of the history of humanity are incredible resources". This was a departure from the language of service, which positions knowledge as "a gift bestowed by those who consider themselves knowledgeable upon those whom they consider to know nothing" (Freire, 1970, p. 72). As I describe in the next chapter, the founders and early educational leaders of the museum viewed museum education through a "charitable" lens—gifting public education students and teachers access to exclusive plots of scientific knowledge to enrich the perceived impoverishment of their lives. Moving to a language of resources ascribes more agency to learners as they can have the opportunity to adapt and use the resources to meet diverse goals and confirm diverse identities. However, like any other resource, it is important that learners not only learn about what the resource affords—what the resource enables them to do—but have the space to think about the different ways they could use and adapt/modify the resource, in the case of teachers, to meet their learning goals and students' needs.

Learning is the creation of culture. Museums and schools have different cultures of learning; therefore, effective connections require building relationships in ways that allow for co-learning and co-development of integrative learning cultures and processes. This also requires knowing historical patterns of learning cultures: recognizing those things on which to build and expand, as well as those things that are sometimes explicit but often hidden yet made visible to be dismantled. Through partnerships, AMNH helped to shape an expansivising space where learning practices shifted and broadened, and relationships with schools, teachers, students, and classrooms were deepened. In the following section, I describe how some of the partnerships between AMNH and schools unfolded.

School–Museum Partnerships for Science Learning

In the 1992/3 Annual Report (1993), the museum's education department wrote as its goal:

> To have an individual come away from his or her visit to the Museum with a greater comprehension of the complex ecological issues confronting the world as it approaches the 21st century...major objectives that underscore the department's systemic change initiatives are improving teacher education in natural history, thereby facilitating an understanding of scientific endeavor."

The use of the word *understanding* in this statement indicates that the Museum as an institution realizes that it must do more than just present content. As a facilitator (rather than a silent teacher or benefactor), the Museum must engage/initiate the learner in the practices of sciences and devise ways for learners to understand the importance of science in their daily lives and in larger global perspectives. One of the ways to engage in this kind of learning was through systematic and sustained partnerships with teachers and schools. The museum began its school–museum partnership initiatives in the latter 20th century. There were three models of partnerships: AMNH in partnership with several schools, one school in partnership with several museums, and a partnership between several science-rich museums and the New York City Department of Education, which includes institutions and schools across the city.

New York City is one of the largest urban districts in the United States. It has a considerable number of schools that are situated in lower-income, immigrant communities and with students from Black and Latinx backgrounds, populations that are grossly underrepresented in STEM. Many schools' science programs needed more licensed science teachers and adequate science teaching and learning resources, such as labs, expendable materials, teaching/demonstration models, and up-to-date media. Professional development at the museum and other science-rich cultural institutions was structured to support science teachers and those assigned to teach science in strengthening science content knowledge in the context of using the museum and other ISIs as resources to support classroom science learning. With the notion of "museum as a resource" in mind, the objects and exhibits would provide science classrooms with opportunities for observing, describing, questioning, and documenting—engaging in practices foundational to scientific inquiry. As precursors to charter schools, the small schools trended in the late 1990s and presented a suitable time for AMNH and NYC public schools to build partnership relations and serve as models for other small schools across the city. The small schools aimed to improve student achievement through themed schools and alternative assessments. Museums offered spaces and resources that allowed educators to design and enact different approaches to learning.

The Museum School was the first partnership and was created with the goal of "creat[ing] a unique inquiry natural history education model for middle school students and their teachers" (Annual Report, 1992/3, p. 63). This initial partnership included 90 sixth-grade students and 18 teachers from a local middle school, the local community school district, and New York City Public Schools. Teachers in this partnership collaborated with museum educators to create units that were museum-based and connected to the learning standards and lesson topics for middle school. Classes in this partnership experienced "weekly museum sessions using constructivist and cooperative learning theories to investigate intricate natural science concepts" (The American Museum of Natural History, 1992/3).

The Museum as a Classroom

While the Museum School was a partnership with AMNH and several schools, The New York City Museum School (NYCMS) was a partnership between one school and several museums and designed to take advantage of the array of museum resources throughout the city. The NYCMS was established in 1996 with a grant from New York State and designed "to take full advantage of museums, both as classrooms and as models of learning environments to equip students to be lifelong learners," according to the co-founder of the school. The partnership with AMNH was meant to strengthen the science portion of the program. Its unique approach was based on a module system that allowed students to spend extended time on a particular topic that allowed in-depth study of science. It provided students the opportunity to have extended time in the museum that allowed them to observe, question, synthesize, and make connections in their learning. One of the collaborating teachers noted that the goal of the NYCMS is to get students excited about learning in museums. He then offered the "Black kids from the projects" as an example of students who would benefit, repeating the deficit-laden altruistic narrative of early museum programming as discussed in Chapter 2.

Although the NYCMS emphasized long-term, museum-based engagement with topics, there was a considerable difference in experience from middle to high school. The middle school modules were nine weeks long and designed around a topic or theme. To explore the themes, students visited the museum for 2.5 hours twice a week to learn in the halls and the museum classroom. The project-based modules were collaboratively planned by a museum and classroom educator. In New York State, high school students are required to take state assessments, colloquially called "The Regents", to graduate from high school. Thus, the high school museum learning was less immersive and focused on learning content relevant to the exam. The assessments conflicted with the mission and pedagogy of the school, as a NYCMS educator noted in personal conversation:

> "Yes, [State tests are a problem], [unfortunately] I teach to the test. *I am morally and philosophically against tests* (said with emphasis) ... [due to the partnership] not enough time is devoted to skills building [for success on the State exams]. Ten periods in the Museum [a week], those are time away from skill building, they have not even four periods of math a week and expected to take the same test that others spend more time."

Therefore, although the school emphasizes alternative assessment, due to the macrostructure [DOE policy] that the school is beholden to, there must be accountability for the mandated standardized tests. This creates a contradiction in the school (and teachers) of wanting to emphasize extended observations and science inquiry in tune with the culture of science in the Museum while having to dedicate time to classroom-based test-taking skills practice for success on the standardized tests.

Urban Advantage

Urban Advantage was initiated in 2004 as a partnership between the Museum, six of the city's science-rich cultural institutions and the New York City Department of Education. An overarching goal of the partnership was to make the resources of the city's science-rich institutions accessible to teachers, students, and their families. The partnership was formed around the Exit Project requirement for middle school graduation. An Exit Project is a long-term science investigation. Although each partnering institution has its own history of working with schools, the Urban Advantage initiative created a collaborative, city-wide structure for supporting urban science education.

The Urban Advantage partnership began during a time when the New York City Department of Education was changing to meet the United States federal *No Child Left Behind* (NCLB) mandates. Initiated in 2002 during the George W. Bush administration, the NCLB was described by Zeus Leonardo (2009) as "the educational cognate of the *Patriot Act* following the terrorist attacks on the World Trade Center in 2001, through its emphasis on nationhood and Americanism" (p. 268). Targeting racialized, immigrant, lower-income students as well as students with disabilities, the NCLB created a militarized culture of schools that emphasized standardized assessments with penalties for teachers and schools not meeting specified criteria. The NCLB did not account for long-standing structural issues that placed the targeted students at the margins but rather assumed a color-blind "fairness" in both foundations and enactments of the national policy (Leonardo, 2009).

New York City's public schools serve large demographics of racialized and/or immigrant students, with many falling near or below the poverty line. To meet federal mandates and not lose funding, literacy and math became the central focus of instruction, relegating science to a secondary position of teaching and learning priorities. Around the same time, the NYC Department of Education (DOE) formed a dedicated science department with a director of science, Dr. Julia Rankin, for the first time in its history. It was also in this climate that AMNH, in partnership with the DOE, aimed to establish a city-wide network to strengthen science teaching and learning in the NYC public schools.

As Dr. Rankin and the new Department of Science were working to reprioritize science instruction in the schools, the Urban Advantage partners were planning the structure and content of the program. I was part of the planning, and we had to rely on our previous work with middle school students and teachers alongside existing Exit Project descriptions and eighth-grade curriculum for initial structuring. Since the Exit Projects emphasized scientific inquiry, the partners thought about how we could authentically use the ways that scientists approach their research and the museum's objects and exhibits in the program's design. Concerned about the NCLB mandates, the partners also considered how learning in science-rich cultural institutions could support math and literacy.

At AMNH, we were able to leverage lessons learned in partnerships with schools and teacher professional development in the design. Through the museum–school partnerships, we learned that extended contact in the institution, both for the teachers and students, was important for learning and supporting long-term science inquiry projects. However, there were still questions about how the project would play out in relation to classrooms because, unlike NYCMS, schools across the city did not have extended time structured into their days, let alone travel time between the school and institutions. While teachers were excited about the initiative, early on, they discussed challenges with integrating field trips and museum-based resources with doing science investigations and Exit Projects in the classroom. The science-rich cultural institution partners realized that to ensure success with the project, teacher expertise was needed in the development and advancement of Urban Advantage.

Evolving Partnership Practices

For successful learning in the partnerships between museums and schools, museum and classroom educators must become familiar with each space's schema, resources, and practices. As a collaborating museum educator mentioned:

> [In the traditional high school] I found teaching to be a solitary experience; you close the door, do my thing in the classroom, and except for the occasional observation I would not have any real feedback. At NYCMS and here [AMNH], I have had the opportunity to do collaborative teaching over an extended period. The single most important thing is the ability to collaborate, not only to plan with, but also to be in the room with another teacher and be able to reflect with another teacher who has been in the classroom with me. It is shaped by AMNH now, I can collaborate with my partner teacher, but I find it easy to seek out opinions from others in the museum who are not working with me. The opportunity for me to collaborate has increased here [at AMNH].

This exchange of schema and resources creates an expansivising space between museums and schools. The museum educator in the partnership noted, "Anytime you are in a collaborative experience, you will pick up ideas; everything is collaborative, and this has played a big role in my pedagogy. Seems like a very important piece, played a significant role in my growth as a teacher." Similarly, in a 2001 interview, the co-founder of the NYCMS school mentioned collaboration as central to the school's mission, "teachers and museum educators work together and share their expertise." This collaboration between teachers serves as a model for students as they collaborate with their peers on projects. It is interesting that it also reflects the working practice of scientists in the museum, as collaboration is an essential element in what they do.

Partnering with teachers and schools helps to make the museum resources more relevant and responsive to supporting city-wide science education efforts. Although the exhibits remain central in teacher learning, they are no longer the starting point of developing teacher learning sessions. The standards and curricula guide the selection of the exhibits and displays that will form the basis of learning interactions. With newer or temporary exhibits, the scientific discoveries and themes are highlighted while also making deliberate connections to the curricular needs of the classroom. This is especially relevant for special and/or temporary exhibits designed around museum research and/or cutting-edge discoveries. These also have the advantage of having the newest and most engaging modes of presentation in terms of digital resources, interactives, and visual/object displays.

School Science Contrasting to Museum Science

Good science teaching and learning is resource-dependent. It requires equipment, tools, and objects, some of them expendable, to model aspects of "real-world' science as practiced beyond K-12 education. As I moved from being a classroom teacher to being a museum educator, I embarked on my own expansivising journey of learning about the schema and resources of the different contexts and how to put them in relation first to student learning and then to teacher learning.

The school where I first taught was a traditional, large, comprehensive high school. With over 4000 students, the school had one wing dedicated to the physical sciences (Earth science, physics, chemistry) and another for the life sciences. There were lab techs who were charged with setting up labs and maintaining the well-stocked supplies. While there were ample science education supplies, the science instruction was didactic and mainly geared towards students passing the standardized tests. The labs had very little inquiry, and the classroom lectures were just that—lectures with demonstrations (demos) of scientific phenomena done by the teacher at the front of the room.

When I started teaching at this school, there were still many students from the adjacent white communities. Many of these students were sequestered in the honors program, with access to upper-level science (chemistry and physics) and even Advanced Placement electives. In other words, in a school where more than half of the students were not white, in an honors class with the teachers' union class size limit of 32, you saw, at maximum, 3 non-white students.

I was the youngest member of the science department with all my colleagues being more than 20 years my senior. I was also one of the very few Black teachers in the school and the only non-white person teaching science. My colleagues were white and mostly former Brooklyn residents who now lived in the suburbs of Long Island. They formed the white flight that left Brooklyn when the Black Caribbean population expanded during the 1970s.

With the shift in community and student demographics, the elective and upper-level classes disappeared. What remained was the bare minimum to meet the required state offerings of high school science. Both teachers and learners were reproducing "simulacra" (Lemke, 1992) of science where the practices, objects, and discourses of science in the classroom were decontextualized from science as practiced in research institutions and lived in students' day-to-day experiences, "They encounter simulacra of science as a process of activity: school laboratory exercises in place of professional investigative practices; efforts to solve problems that have no real contexts, no real parameters, no realistic complications; study of examples and are idealized, oversimplified and decontextualized" (Lemke, 1992). Emphasis was placed on knowing scientific facts rather than engaging students in scientific knowledge production by fostering their curiosities and connecting them to relevance.

When I moved to a different school, it was at the beginning of the small school movement in the city. Claiming to attenuate the anonymity found in the larger high schools, the smaller schools were conceived to offer students more intimate and safer learning environments. The small school in the small building was in the South Bronx—all the students represented urban, underserved groups in multiple intersectional ways. Here, science was even less present. There were no dedicated lab spaces for science, and the science available was not enough to supply the entire school. I did not have access to microscopes or even the simplest glassware for students to become familiar with the tools of science. We were a very small team of science teachers and made use of the 99-cent stores to supplement science learning in the classroom.

In both schools, I was lucky enough to find ways to engage students in science learning outside of the simulacra of the classroom. In the Brooklyn school, I was able to connect with an outdoor education organization, New York City Outward Bound, and co-create expeditionary learning experiences that were socio-environmentally focused. In this class, students went to the museum and the local parks and seashores and even conducted long-term engagements of plant diversity in the school courtyard. Students were able to explore Earth science and the life sciences through the lens of place-based education. Although there was a curriculum, students were able to do more individual and group explorations of their interests as they emerged from their observations. The Bronx school was located across the street from the zoo and adjacent to a small city park with a waterfall. I could take students to the park, where we took water samples and even found a planaria! The school also had a relationship with the zoo, where students were able to get credit for an after-school animal studies class. Twice a week, I walked with a group of students through the zoo and to the classroom, where they started their behind-the-scenes activities. In both schools, teaching elective classes allowed me to create these kinds of expansive learning experiences. When I taught the classes that ended in the state-mandated assessments, time only allowed me to cover the curriculum and the required, recipe-like lab activities.

In the elective classes, the integration of out-of-classroom/school activities enabled me to move students beyond the simulacra to a more lived and actively practiced science. These opportunities also allowed me to expand my knowledge of my students as learners and as unique humans more than I did in the formal classroom. The learning was engaging, cooperative, immediately relevant and inspired longer-term investigations. Also, importantly, we were able to get to know each other in humanizing ways through the different conversations that emerged between science activities. For example, informal science learning often requires walking together—whether to get to and from the site or to different exhibits within the site—walking together is the opportunity for students and teachers to engage in personal learning conversations that are important in building trust and resonance that create safer science teaching and learning spaces. Thus, as a teacher, these informal learning environments were also a source of professional learning in many ways, as they involved engaging the multiple dimensions of being a science teacher, such as knowledge of science and how to "do" science, knowledge of the ways that diverse students learn science, knowledge of the self-who teaches vis-à-vis their students. For me, museums were emergent laboratories for my teacher learning. Being a teacher educator in an ISI setting allowed me to structure learning for teachers that also engaged the multiple dimensions of being a science teacher.

My teacher learning has been ongoing, and my experiences of teaching in informal science settings have allowed me to embrace the emergence and value of dialogic interactions between students as learning activities in my own practice (now as a professor in post-secondary education). Emphasizing teacher agency, the following sections describe how teachers learn to teach within and across contexts.

Teaching and Learning to Teach

Teachers are always in the process of learning to teach as they encounter new students, resources, and challenges within the constantly shifting, socially dynamic contexts of science teaching and learning. It is a continuous process that happens both in the moment, across time, and in different contexts. Teachers gain expertise as they gain experience; however, the constantly changing mandates, as well as students, classrooms, and access to resources, require that teachers gain Spielraum—the ability to appropriately respond to the constantly unfolding social life of teaching and learning (Adams & Gupta, 2017). Teachers develop Spielraum not only in the experience of teaching but, more importantly, through opportunities to reflect on teaching experiences and generate teaching imaginations with other educators. Teaching imaginations involve the ability to create teaching scenario possibilities of "what-ifs" that could then lead to actual enactments in the classroom. Teacher education that supports Spielraum structures and integrates spaces for teachers to reflect on their learning and discuss how the teaching could transfer into practice with other teachers and educators. These dialogues with others allow for sharing challenges and ideas and brainstorming possibilities for action.

When I facilitated workshops for science teachers in the museum, I got to see first-hand their excitement and sense of inspiration from learning in the halls and with museum educators and scientists. We emphasize engaging teachers as learners; we modeled facilitation and immersed teachers in hands-on activities both within the museum and outside on the museum grounds to allow them to embody the experiences they could then enact with students. I often heard, "I can't wait to bring my students here to do this!" For in-service teachers, their reflections are based on what they already know is possible in the classroom and their hopes for more opportunities to engage their students in the excitement of science. For preservice teachers, this might be more of a reflection and imagination based on their own experiences and knowledge about the notion of "being a teacher." As teachers gain classroom experiences, the teacher learning happens during the enactment of teaching, which eventually becomes embodied and in-the-moment enactments that culminate in patterns of action or practices. For example, a teacher will know how to sense tacit changes in student energy and engagement and shift practices in relation to this.

Language Structures Practices

Teachers make pedagogical choices and enact practices based on the meanings that they make of their teaching-learning experiences. In a survey about teachers' informal learning experiences, I applied a key word in context (KWIC) (Leech & Onwuegbuzie, 2008 in Adams & McCullough, 2021) approach to gain some insights into how they interpreted "informal science." I examined data for instances where teachers used "informal" and "informal science," the words and phrases that were associated with those words and how their discussions revealed evidence of practices. Unsurprisingly, teachers associated the term informal science education with field trips. With this, they mentioned specific places they visited with students or where they participated in courses or professional development. Teachers also held informal science in a strict dichotomy with formal science. Terms like "rigid" and "structured" were applied to formal science learning, whereas "dynamic" or "flexible" were ascribed to informal. Informal science was described as more "authentic," allowing students to engage in practices more aligned with how scientists conduct research. A theme of unpredictability also emerged, similar to practicing scientific research teachers who expressed being unable to control the outcome of informal science learning but viewed this positively. In general, authentic science experiences where students collect real data and open-ended labs were associated with informal science learning. Also related to informal science learning were student choice, inquiry-based learning, and going beyond the given curriculum.

ISIs and Teacher Education

Transformations in science teacher education have focused on embedding more inquiry-based, culturally sustaining, and authentic science learning experiences in teacher learning. Good teachers know good teaching, and those in my studies have recognized that effective and quality science teaching requires a continuum between classroom and out-of-classroom experiences. This engagement with informal learning is critical in equitable science learning opportunities and engaging learners with diverse abilities and backgrounds (Adams, DeFelice & McCullough, 2022; Adams, 2019).

In North America, this is especially critical in urban areas where schools educate large percentages of racialized students who have been historically marginalized from science majors and careers and are more likely to be lacking in qualified science teachers and practical science teaching and learning resources, such as adequate labs and updated experimental equipment (Darling-Hammond & Sykes, 2003; Coble, Smith & Berry, 2009). Connecting ISE teacher preparation to topics taught in the curriculum allows teachers to build a deeper understanding of topics and facilitate experiences that access a more comprehensive range of learners. ISIs such as natural history museums, zoos, and botanical gardens have a variety of objects on public display that complement classroom science. With access to authentic scientific objects and displays of scientific phenomena, science-rich cultural institutions have the potential to provide valuable learning experiences and professional education for science teachers. Early iterations of museum-based science learning at AMNH were done with Brooklyn College, CUNY.

In an early study of a collaborative course between a natural history museum and a four-year public university, researchers, and museum educators Dhingra, Miele, Macdonald, and Powell (2001) reported that teachers experienced a positive difference between learning in a college classroom and learning in a museum. Teachers described the learning experience as being experiential and engaging. The researchers noted that this could translate into creating more of an interactive learning experience for their students (Dhingra et al., 2001). This established the idea that teacher learning in ISIs could lead to improved science teaching and learning.

In the workshops I planned and facilitated at AMNH, teachers learned science from viewing objects in the exhibition halls and closely investigating related objects in the classroom. Workshop co-facilitators, who include scientists, museum educators and/or classroom educators, modeled how to incorporate the museum's resources into various science topics. For example, a facilitator would use a diorama in the Milstein Hall of Ocean Life to discuss ocean ecosystems while modeling the use of the diorama as a teaching resource at the same time. The workshop participants were afforded the opportunity to learn about science content (such as animal and plant adaptations) and experience effective use of the museum's resources. The co-teaching of the workshops by scientists and educators helped to strengthen teachers' content knowledge. From the scientists, teachers discussed and learned about real-life applications of the inquiry process and the rationale behind the display of particular objects or specimens in exhibits. They also learned how and what scientists learn from studying similar objects in the field and lab. Teachers taught the scientists what aspects of their work were relevant to classroom curricula and how to communicate their research in accessible and engaging ways. This exemplified the collaborative nature of teacher learning and demonstrated how museum and classroom educators and scientists learn from each other towards the goal of improving science education.

Designing Museum Learning for Educators

As manager of curriculum initiatives, I coordinated the collaborative designing and facilitation of the workshops. We knew that teachers enjoyed participating in the workshops not only for the engaging experiences in the classroom but also for the resources we often provided them to bring back to the classroom. The supplementing materials ranged from compact discs with the Space Show and other scientific media produced by the museum, hall guides, and sometimes even objects, such as rock samples and magnifying glasses for classroom use. These workshops were designed around the themes found in the museum and connected with the school curriculum, usually topics in biology or Earth Science because those were the topics of the standardized tests, and with the intention of teachers being able to use these halls on field trips with their students to supplement.

The general format of workshops started with an introduction in a classroom space within the museum. This session introduces the relevant halls, including discussions about the design of the newer halls, scientific research in the museum related to the content of the halls (many times having a museum scientist speak about their work) and a classroom activity to prepare for doing inquiry in the halls. As observation and questioning are key practices used to engage learners with exhibits and objects in the halls, the table activity emphasized initial open-ended questions and observation prompts to engage learners in careful and critical observations.

As we moved to the halls, we framed the visits as expeditions. Expeditions were integral to the museum's history and the source of many historic objects and specimens, such as the taxidermy in the dioramas, and cultural objects, including sacred meteorites. We were sure to be at the forefront of the controversies around collecting while explaining that science has learned a lot from objects. Once in the halls, we would facilitate activities that used questions and observations to engage teachers with the exhibits. We would structure questions ranging from asking teachers to document what they see to asking them to make connections between different observations conducted across exhibits or halls. The questions would aim to make static displays, like dioramas, appear more dynamic by asking teachers to make inferences about the activity of organisms in the diorama. For example, we would ask them to think about the diorama as a snapshot of a moment, make inferences about what happened immediately before, and predict what might happen afterward. This would prompt them to look for cues in the diorama and make connections across different observations. We would also ask questions about relationships, for example, "What do you think are the relationships between the different organisms you see? What are the relationships between the organisms and the environment depicted? What do you think might be missing? What message do you think the curator of this display was trying to convey? These kinds of questions allow the teachers-as-learners to observe the science embedded in the exhibit and begin adopting a critical stance towards looking and visual literacy. We would also structure activities that required them to gather information from different halls to address more prominent themes or inquiry questions. Returning to the classroom space provides time to debrief and discuss their engagements as teacher-learners and transference into practice in the classroom and with students in the museum.

These workshops focused on teaching teachers how to use the museum's resources both in the museum and the classroom. We emphasized a "pre-during-post" format where the museum trip served as an integral component of a particular topic or unit. The pre were activities that would both engage students with the content and familiarize them with the practices they would do in the museum. For example, since we emphasized making careful observations of objects, asking questions about the objects, and making inferences that would lead to further investigation, we encouraged teachers to use objects and observation instruments (such as magnifying glasses and science notebooks) to engage students in these practices before their trip to the museum. The purpose was to reduce the levels of distraction a student might experience in the museum—both being in a novel and stimulating environment and engaging in a new practice. The during included examples of activities teachers could use in the halls with their students. Some of these will be described in subsequent chapters but almost always included observations, posing questions, and gathering information. We always emphasized that students were not expected to get all their answers at the museum but to use the trip as a way to spark further interest in the subject. The post activities were aimed at having students bring their observations and "data" back to the classroom to connect with the curriculum and other topics in their science. During the workshop, we also introduced online resources curated by the museum to support extending students' observations and learning into the classroom. We found that this pre-during-post model was a salient way of both engaging teachers as learners and providing a model for them to plan focused field trips with their students.

Learning, Teaching, and Learning to Teach Science

Teaching, learning to teach, and identity exist as collateral discursive practices in that each influences the other, and they all evolve and transform together. This is not without the influence of teachers' social identities, which also subject them to the racial, gendered, socioeconomic, ableist, and other discourses that exist in institutions and society. When teachers learn to teach, they reproduce identities as being certain kinds of teachers (Gee, 2000–2001), which depends on a complexity of factors. In teaching science, the interactions with physical and material resources become more important as effective science teaching requires access to objects of science, including lab equipment (even basic magnifying glasses and measuring instruments), specimens, objects, as well as well-ventilated spaces and access perishable supplies to do laboratory exercises. Therefore, the resources that science teachers have access to and know how to use will also influence their teacher identities. The institutions and colleagues they associate and identify with and how they enact teaching in the classroom—how they design learning environments and make resources available to their learners- are additional important factors shaping science teacher identities.

Teacher Agency and Identity in Relation to Self, Place, Practice

A framework that posits a dialogical relationship between identity, agency, and learning to teach is salient for understanding the relationship between informal science learning and science teacher identity. Through the process of learning to teach, a teacher develops an identity in relation to both the learning cultures and contexts and interacts with others in these contexts. Teachers enter professional learning with notions of what it means to be a teacher and shape these notions based on the learning cultures they encounter in their teacher learning and their enactments of teaching and in relation to others (students, colleagues, administrators, etc.) throughout a career. Through this process, they develop agency in teaching: learning and knowing which resources to use, when, and why to teach the students as social agents in their classrooms. Defining teacher identity as "the ways in which a teacher represents herself through her views, orientations, attitudes, emotions, understandings, and knowledge and beliefs about science teaching and learning" (Avraamidou, 2014, p. 826) allows us to move beyond what is learned to focus more on the learning cultures in which the teacher-as-learner engages and allows us to ask questions about who a teacher is and what does this means in terms of how she teaches (Beauchamp & Thomas, 2011) and how this is a function of multiple social and material interactions. In this sense, "identities are the part of self that are defined by the different positions we hold in society" (Varelas, 2012, p. 3); thus, teacher identity is not an end-product but rather an ongoing process of teaching and learning about self, others, resources, places, and practices in different contexts, and who they are in relation to others.

Agency is the belief that the self is capable of effective science teaching. This means making the right pedagogical decisions, adapting, and using different resources to meet those pedagogical decisions, having confidence in science content knowledge, and engaging students in science learning. Through the process of learning, one gains capacity in these skills and begins to develop an identity associated with competence in those skills. Depending on the learning contexts and one's self-perception in relation to others, one starts to define oneself as a "kind" of teacher, whether it is inquiry-based, hands-on, fun, hard, strict, etc. Important to developing agency and a corresponding identity is being able to access appropriate affordances available to be an effective teacher. Affordances include physical and intellectual resources, practices, social and professional networks, and other resources that shape and enable teaching and learning. Through the agency, teachers appropriate and adapt affordances available, or, in a polyphonic bricolage (Schmidt, 2008), create new resources from existing ones to create or maintain a particular teaching identity. Agency allows one to transform how one uses affordances within and across settings to expand and transform science teaching and the learning opportunities available.

Subjectivity is identity-in-relation-to. It is the relational aspect of identity that is constantly shifting and changing in relation to contexts that one encounters, "the conscious and unconscious thoughts and emotions of the individual, her sense of herself, and her ways of understanding her relation to the world" (Weedon in Jackson & Mazzei, p. 52). It is impossible to separate identity and subjectivity—they are two sides of the same coin. Identity is being a kind of person (Gee, 2000–2001); the subjectivity of identity is "not stable, but is constructed in relationships in others and in everyday practices—an ongoing process of "becoming" rather than merely "being" in the world" (p. 53). In Stetsenko's (2011) notion of Transformative Activist Stance, we are always in the process of becoming change agents; we are changing our world as we are learning about our world and our place in it. The subjective identity shifts among the discursive fields or learning cultures that create it; as such, becoming and being a teacher shifts with the learning culture they encounter as they create their identities as professional science teachers. Teachers develop agency as they encounter different learning cultures that include resources, ideas, historical structures, policies, normalizations, and other identity-subjects and learn associated practices.

As sites for teacher learning, ISIs have the resources that could support science teacher identity and agency development in several salient ways. Teachers learn:

(a) Pedagogical approaches that emphasize observation and inquiry, which are readily transferable to different contexts;
(b) Accessing the affordances of visual/object-based culture in learning;
(c) Approaches that extend learning outside of the classroom and into the community;
(d) Approaches that allow them to observe and respond to student learning in context.

The salience of informal science learning for classroom teachers is the idea of utilizing museums and science-rich cultural institutions as resources that support science learning. As science is a material-dependent subject, effective science teaching requires the use of objects and tools of science in the labs and classrooms, and teachers in schools with inadequate resources are compelled to augment science learning by looking beyond the classroom. For many teachers, this means field trips to and/or professional learning opportunities that allow them to learn about using museums and other science-rich cultural institutions to enrich science teaching and learning. Teachers make pedagogical choices and enact different practices based on what they learn and experience, what they know and believe about their students, and the identity they aim to create and maintain as teachers.

When schools lack adequate science education tools and resources, ISI pedagogical approaches do not make up for this dearth, but they can augment it. Teachers and students can connect with up-to-date scientific advances and discoveries and access resources in their communities to engage in meaningful science teaching and learning. ISE could provide a framework for teachers and students to question and investigate their worlds together while building scientific knowledge that would afford them more agency over their own learning and the futures they envision for themselves and their communities.

Looking Forward

Chapter One introduces some of the historical and theoretical underpinnings of the ISE field and how it relates to science teacher education. It presents the ways that partnerships and collaborations between formal and informal science can improve learning opportunities for teachers and also expand affordances for science learning in the classroom. Chapter One also introduces the Urban Advantage partnership that serves as the context for the research described in Chapters Three to Five. Many museums and other science-rich cultural institutions have long histories and corresponding ideologies that help to shape the affordances therein. In order to transform the practices of museum–school collaborations, it is important to elucidate this history, including how it has influenced and shaped the science and science learning promoted by the cultural institutions. Chapter Two addresses this by describing some of the underlying history and corresponding tensions often hidden when considering museums and other informal institutions as contexts for science learning. Just as the history of schools and schooling is learned in teacher education and education research, other institutions integral to teaching and learning should receive the same treatment. This is especially critical when discussing issues of equity, justice, and human agency in learning.

The teacher-learning/learning-to-teach process is a central theme in this book. As such, Chapter Three presents the theoretical underpinnings of teacher learning-to-teach and teaching across formal/informal settings. Using the relationship between ISE and schools/classrooms as contexts, I describe how learning and leveraging the affordances of each shape teacher identity and agency, expanding opportunities for student learning. Urban Advantage was an initiative developed to help expand affordances for science learning across a large metropolitan district through access to science-rich cultural institutions. Teacher learning was central to this initiative.

In Chapter Four, the book highlights a group of teachers, the Urban Advantage Lead Teachers, who helped to create an expansivising space between the museum/ISI and schools. It will highlight some of the challenges and successes they experienced as well as the classroom-as-artifact that they developed to demonstrate the learning culture developed in the expansivising space. These classrooms also provided spaces to facilitate discussions around developing practices of inquiry-object-based learning, one of the central practices learned in the Urban Advantage program. Through their discussions and interactions with their artifact, the Lead Teachers expanded their agency within the school and cultural institutions in various ways. Violet is one of the Lead Teachers who used Urban Advantage to expand the affordances and transform practices within her school for science teaching and learning despite the silencing of science in favor of literacy and math. Chapter Five presents a narrative of how she worked to make science teaching and learning more central in her school despite administrative and collegial challenges. The final chapter focuses on the Informal Learning Environments and Teacher Education for STEM (ILETES) project and describes the collaborating teachers who created and enacted their meaning-making around ISE to enact meaningful science learning for their students. Their dialogues about science teaching and learning over three years allowed them to build a common vocabulary while developing a range of practices reflective of their identities as teachers and the science learning they imagined for their students, especially their racialized and marginalized learners. It highlights the importance of dialogic spaces for teachers where they engage in ongoing cycles of learning affordances and enactments and reflect on those enactments vis-à-vis the realities of the classroom to develop agentic teaching identities and equity-centered repertoires of practices, thus expanding opportunities for students' science learning and engagement.

Notes

1 Synonym of fresh or dope, a 90s term that refers to being cool, up-to-date in style and fashion and usually would not want to "hang out" with an adult, nerdy teacher during their free time.

References

Adams, J. D., DeFelice, A., & McCullough, S. (2022). Teacher-learning, meaning-making, and integrating ISE practices in diverse urban classrooms. *Connected Science Learning*.

Adams, J. D., & McCullough, S. (2021). Inquiry and learning in informal settings. In Clark A. Chinn & Ravit Golan Duncan (Eds.), *International handbook of inquiry and learning* (pp. 358–370). Routledge: New York.

Adams, J. D. (2019). WhatsApp with science? Emergent CrossActionSpaces for communication and collaboration practices in an urban science classroom. In *Emergent practices and material conditions in learning and teaching with technologies* (pp. 107–125). Cham: Springer.

Adams, J. D., & Gupta, P. (2017). Informal science institutions and learning to teach: An examination of identity, agency, and affordances. *Journal of Research in Science Teaching*, 54(1), 121–138.

Gupta, P., & Adams, J. D. (2012). Museum–university partnerships for preservice science education. In *Second international handbook of science education* (pp. 1147–1162). Dordrecht: Springer.

American Museum of Natural History. (1992/3). Annual Report. *AMNH Digital Library*.

Avraamidou, L. (2014). Developing a reform-minded science teaching identity: The role of informal science environments. *Journal of Science Teacher Education*, 25(7), 823–843.

Beauchamp, C., & Thomas, L. (2011). New teachers' identity shifts at the boundary of teacher education and initial practice. *International Journal of Educational Research*, 50(1), 6–13.

Coble, C. R., Smith, T. M., & Berry, B. (2009). The recruitment and retention of science teachers. In *The continuum of secondary science teacher preparation* (pp. 1–21). Brill.

Darling-Hammond, L., & Sykes, G. (2003). Wanted, a national teacher supply policy for education: The right way to meet the "highly qualified teacher" challenge. *Education Policy Analysis Archives*, 11, 33–33.

Dhingra, K., Miele, E., Macdonald, M., & Powell, W. (2001). *Museum-College-School: A collaborative model for science teacher preparation*. Paper presented at AERA, Seattle, Washington.

Dierking, L., Falk, J., Rennie, L., Anderson, D., & Ellenbogen, K. (2003). Policy statement of the "Informal Science Education" Ad Hoc Committee. *Journal of Research in Science Teaching*, 40, 108–111.

Falk, J. H., & Dierking, L. D. (2004). The contextual model of learning. In *Reinventing the museum: Historical and contemporary perspectives on the paradigm shift* (pp. 139–142). AltaMira Press.

Falk, J. H., & Dierking, L. D. (2002). *Lessons without limit: How free-choice learning is transforming education*. Rowman Altamira.

Gee, J. P. (2000/1). Identity as an analytic lens for research in education. *Review of Research in Education*, 25(1), 99–125.

Jackson, A. Y., & Mazzei, L. A. (2011). *Thinking with theory in qualitative research: Viewing data across multiple perspectives*. New York: Routledge.

Leinhardt, G. (1997). Instructional explanations in history. *International Journal of Educational Research*, 27(3), 221–232.

Lemke, J. L. (2001). Articulating communities: Sociocultural perspectives on science education. *Journal of Research in Science Teaching*, 38, 296–316.

Leonardo, Z. (2009). *Race, whiteness, and education*. Routledge.

Rowe, S. (2002). The role of objects in active, distributed meaning-making. In G. Paris (Ed.), *Perspectives on object-centered learning in museums* (pp. 19–36). Mahwah, NJ: Erlbaum Associates.

Sadoski, M., Paivio, A., & Goetz, E. T. (1991). Commentary: A critique of schema theory in reading and a dual coding alternative. *Reading research quarterly*, 463-484.

Schmidt, B. (2008). The many voices of Caribbean culture in New York City. In H. Henke & K.-H. Magister (Eds.), *Constructing vernacular culture in the trans-Caribbean* (pp. 23–42). New York: Lexington Books.

Stetsenko, A. (2011). From relational ontology to transformative activist stance on development and learning: Expanding Vygotsky's (CHAT) project. In *Marxism and education* (pp. 165–192). New York: Palgrave Macmillan.

Varelas, M. (2012). Introduction: Identity research as a tool for developing a feeling for the learner. In M. Varelas (Ed.), *Identity construction and science education research* (pp. 1–6). Brill Sense.

· 2 ·

AN ARTIFACT OF SETTLER COLONIALISM

Race and Historical Displays in the Museum

| Henry Fairfield Osborn—Paleontologist, eugenist [sic], and museum president from 1908 to 1933—oversaw the installation of the first dioramas. The words on his bust celebrate him as a godlike reanimator of the past: "For him, dry bones came to life and giant forms of ages past rejoined the pageant of the living." But these are bones pulled from colonized land, and they are witness to histories older and deeper than Osborn. His belief in the racial superiority of Nordic peoples shaped the arrangement of the museum's collections and they have not been substantially altered over the last century (*Decolonize This Place*, 2016). | You have little idea in walking through these halls what labor they have involved, what sacrifices men have made and are making for them today in all parts of the world, how much the workers in this Museum are imbued with what may be called the spirit of the institution—the desire to extend the call and vision of nature (Osborne, 1927, p. 271). |

The words on the left are the retelling of the legacy of the fourth president of the American Museum of Natural History, Henry Fairfield Osborn. He was a eugenicist and wrote the foreword to *The Passing of the Great Race* by Madison Grant, a long-past president of the New York Zoological Society and Trustee of AMNH. Grant was an unabashed racist and referred to New York City as *"cloaca gentium,"* which "will produce many amazing racial hybrids and some ethnic horrors that will be beyond the powers of future anthropologists to unravel" (Hartman, 2016). His book inspired Hitler, and his sentiments underlie much of the nationalism and fear-of-a-Black-planet mongering of the 45th presidential administration of the United States. In his preface, Osborn wrote,

> "If I were asked: What is the greatest danger which threatens the American republic to-day? I would certainly reply: The gradual dying out among *our* (emphasis mine's) people of those hereditary traits through which the principles of our religious, political, and social foundations were laid down and their insidious replacement by traits of less noble character."

He equated the conservation of the [white] race ("*our* people") with patriotism or love of America. As the president of the museum for 25 years, Osborne was undoubtedly influential in many of the museum's early initiatives, including the design of exhibits and educational programs; it behooves the question of in what ways are Osborn's racist views imbued in the "spirit of the institution" which he played a key role in defining.

In 1921, with Osborn at the helm, AMNH hosted the Second International Congress on the pseudo-science of eugenics. At the time, the study of eugenics was a developing area of research with the goal of "improv[ing] the natural, physical, mental, and temperamental qualities of the human family" (Norrgard, 2008). The actual document of the Eugenics Record Office contains such language in its "principle business" as "to secure social ideals as will facilitate the mating of the fittest" while suggesting controlling coupling amongst young people to ensure that governments have the tools to secure racial progress by "encouraging the reproduction of the 'best blood,' and discouraging or preventing the reproduction of its worse strains," and to encourage every "intelligent and patriotic family to establish a Eugenical Family Archive" for the preservation of genealogical and biographical material that would "aid greatly the practical application of eugenics" (E.R.O., 1927).

As an emerging field, eugenics drew much criticism, such as having no scientific base, offering extreme recommendations for human reproduction and that it was a "gigantic joke." In his retrospective of the museum's role in this movement, Rob DeSalle (2021–2022) noted that "for eugenics to become a full-fledged discipline, the public perception of eugenics needed to be corrected...the organizing committee reasoned that AMNH, because of its educational and exhibition prowess, was the be the perfect place to do it." The scientific authority of the museum would provide the venue that would serve to rarify this emerging, dangerous area of study. Osborn recruited prominent scientists to present papers at the Congress, but several declined, including anthropologist Franz Boas, citing it as a "dangerous sword that may turn its edge against those who rely on its strength" (DeSalle, 2021–2022). There was a section of the meeting on "eugenics and the family" that advocated for the sterilization of prisoners and people with mental illnesses, producing sterilization laws across North America that targeted Indigenous, Black, and poor communities as well as those who were imprisoned or institutionalized with mental illnesses. There was a section that attempted to "scientifically" establish white supremacy with exhibits that covered topics around physical beauty, mental fatigue, and fetal differences in relation to race, including one entitled "Display of the rising tide of color against white world-supremacy" which unironically captures the political ideals that seem to have been recentered in the 21st century. As a person who would have been targeted for elimination by this movement, it is disturbing to learn about this aspect of the history of a museum I love, and which has been a significant part of my life from childhood to the present. Learning this history gives me a particular context and lens to consider the museum's historical underpinnings, including some of the discomfort I felt with some of the objects, art, and displays before I learned of this troubling past. This knowledge must be revealed and considered as we imagine a future of AMNH and museums more broadly as spaces of science learning and communication, community engagement, and important scientific research and innovation. As I will later describe, AMNH has also been taking steps to reconcile with this past.

The American Museum of Natural History is an iconic institution that attracts visitors from all over the globe to view infamous exhibits and objects. However, like many other natural history museums, it also represents an artifact of settler colonialism. Settler colonialism is defined by the erasure of Indigenous people through land theft, genocide, and broken treaties. This also extends to European control of the ecological resources on Indigenous lands, including the extraction and manipulation of materials and organisms (humans and nonhumans alike) for the capital benefit of the settlers and empire. North American museums were designed with underlying settler-colonial ideologies. Although the communicative emphasis is on public education and scientific literacy, the hidden agenda points to the maintenance of a dominant, white-colonial worldview. The ideologies of the founding fathers (who were all affluent, white, cis men) are imbued throughout the museum, and much of the unexamined coloniality remains on display. It is imperative that we interrogate museum histories and practices in ways that allow us to problematize race and racism in these institutions as we move towards antiracist and decolonizing practices. Scholar-activist Porchia Moore (2021) offers salient questions we could ask as we do this work, such as how can we not account for the impacts of the pseudo-science of eugenics and its impact on collections and knowledge in our natural history museums? How should we consider museum best practices, codes of conduct, missions, and standards be considered when the field is overwhelmingly white and does not account for Black and Indigenous ways of knowing? In what ways is implicit bias present when decision-making entities are overwhelmingly white (i.e., curators, board members, high-level administrators, etc.)? These questions are crucial to shaping a historiography as we advance decolonizing, antiracist, and humanizing practices in museums and related learning in museums.

The Roosevelt Statue

Up until 2022, adorning the entrance of the Museum was a statue of Theodore Roosevelt on horseback in oxidized bronze (Figure 2.1). An African and an Indigenous American man flanked him, both on foot. Proudly mounted on his horse, his back is slightly turned towards the African man, who is wearing almost nothing but a downcast gaze towards Central Park. The Indigenous American man—wearing a grand headdress and draped in a blanket—gazes into the park. The figures represent stolen land and humanity, and the design of the statue renders the two men powerless and subjects to the white man on the horse.

This statue aptly represents what Sylvia Wynter refers to as conserving the descriptive statement that positions the European man as the normalized human who "[claimed] "normal" human status by distancing themselves from the group that is still made to occupy the nadir, ["n*"] rung of being human within the terms of our present ethnoclass Man's overrepresentation of its "descriptive statement" as if it were that of the human itself" (Wynter, 2003, p. 262). In this statue, Roosevelt looms over the Black and Indigenous men, both of whom have been dysselected from the category of human, forming the perfectly conserving hierarchical pyramid. Dysselection is Wynter's keyword that references Darwin's evolutionary theory (natural selection) but in a way that legitimizes those who were not "selected" as exemplars of "Man" (Adams & Weinstein, 2020).

In advancing his National Park agenda, Roosevelt was responsible for the erasure of Indigenous people and rural poor from 230 million acres of

AN ARTIFACT OF SETTLER COLONIALISM

Figure 2.1. The Theodore Roosevelt statue displays racist, eugenicist notions of racial hierarchy in bronze. Roosevelt represents the pinnacle of Man with both African and Indigenous humans dysselected from that category.

public land, "These parklands were first and foremost founded as sanctuaries for Anglo-Saxon gentlemen," i.e., places where men of high standing could go and hunt, photograph, explore, and name after themselves (Noisecat, 2019). This is the extension of the call for the vision of nature, a nature that is devoid of Indigenous people and any *cloaca gentium* who dare to tread. It is no coincidence that this statue gazed over Central Park, stolen Lenape land and where the later Black settlement known as Seneca Village was seized through eminent domain for the development of the park.

This statue was a focal point of the Decolonize This Place (DTP) movement. Since 2016, activists gathered at this statue, calling for both its removal and the renaming of Columbus Day to Indigenous Peoples Day. As of the writing of this book, the statute has been removed (Treisman, 2022). DTP also facilitates an Anti-Columbus Day tour where they highlight some of the problematic exhibits on the second floor, where most of the "people" or anthropological halls are located. As they describe in the tour pamphlet, the museum is "frozen in time, bound by nineteenth-century racial classifications that designated human populations as 'primitive' or 'civilized'. Generations of curators have continued this racists legacy, and millions of visitors are invited to take them for granted" (DTP, 2016). While Osborn described the "sacrifices men have made," and, of course, he is referring to the white and mostly male scientists and naturalists of his time, the DTP movement lays bare the authentic and forgotten sacrifices including animals "shot down, stuffed, and hauled to the museum for display," "people exhibited as if they have no history of their own," the exoticization and flattening of Islamic women and African peoples and the contradiction of the mass killing of birds and other nonhuman organisms in the name of conservation. All these histories beckon museum visitors to view these vintage halls with different lenses—ones that question whose history is being told and put on display for whose consumption and the subjugating and colonial ideologies that shaped the former two. These fixed displays without context also serve to fix ideologies.

Reflecting on Representation

When I worked at the museum and well before I had the language and tools to critically unpack the historical context of the museum, I found certain elements in the museum troubling, for example, the murals in the entryway that depict great scenes of great European men on horses and foot collecting animal, mineral, and vegetable treasures to display the wealth of the worlds and peoples that they have conquered in this great building. These murals represent the greatness of Western civilization, with stories of great white men—science and culture—told, interpreted, and displayed from their point of view (Levin, 2002). While many find this hall inviting because of the expanse and natural light, I am disconcerted by the depiction of people of color in subordination to the white explorers on horseback. As an educator, I often wondered if Black, Latinx, and Indigenous students and teachers also saw themselves in subordination on these walls. I also wondered if they saw the history they have learned about themselves in school reinforced on this great wall. Sometimes, when I looked at the statue and murals, I felt a twinge of pain in the wealth and culture that was stolen from many peoples and appropriated for the advancement of white supremacy and generational wealth. Perhaps some people sense this same discomfort when visiting the museum on their own. In my "web of reality," the meaning that I construct from the statue and from these murals is that to the colonial Western mind, the "other" cultures seem only to be as great as the exotica that they produce in the service of colonial advancement (Kincheloe, 2001).

> As they forcibly extracted resources from African peoples, so-called naturalists and explorers like Carl Akeley collected a variety of animals across the continent. Their activities led to species' endangerment, the flourishing of the fur and ivory trade, and widespread deforestation. Akeley was a hunting companion of Teddy Roosevelt and intimate of Belgian King Albert I, who succeeded King Leopold II as the arch-colonizer of the Congo. If you roll up this flyer and put it to your eye, you are looking down the scope of Carl Akeley's gun. Each of the dioramas you have visited is a snapshot of domination: remember this when you look through the glass to see a human on display. (DTP, 2016)

The AMNH contains a wealth of cultural and scientific knowledge both in the halls and behind the scenes. But, like many so-called great things of our nation, there are contentious historical issues that require reconciliation. When classes visit the classic diorama halls, we (museum educators and explainers) are quick to point out that the taxidermic specimens displayed were from a time when it was okay to kill and collect animals for sport. Theodore Roosevelt was known for his fondness of hunting. We explain that while this degree of collecting is no longer practiced, we were able to gain a wealth of knowledge about the diversity of life on Earth from this past practice,

The newer Hall of Biodiversity displays the range of this wealth—the Spectrum of Life wall exhibits representative specimens from each phylum of life on Earth. Individual items, which number hundreds of thousands, are stored as a part of the Museum's collection in their respective scientific departments. While some people are still upset by the collecting practices of the past, many are comforted in knowing that this does not happen (at least to the extent that it did back then) in the present climate of conservation and preservation.

While the murals and cultural halls also contain a wealth of information about the past and important lessons for us to learn today, people take these halls personally. The statue of Theodore Roosevelt on horseback offended me; it felt like a gut punch to my humanity each time I looked at it, so much so that I averted my eyes when I was compelled to pass it. This statue and the entryway murals represent the ideology on which the museum was founded, coloniality, white racial supremacy and power over Black, Indigenous, and other people of color through theft of land, labor, and resources. These statues put all of us on display and serve as symbols that reinforce racial hierarchies and treat oppressors as heroes. Seeing our ancestors displayed in undignified ways vis-à-vis the white man on horseback is both painful and affronting. We can explain the past of the scientific halls in ways that are acceptable to folks today. How do we make visible the ideologies that form the foundation of the ways these cultural artifacts are displayed and read? While it is significant that the Roosevelt statue has been removed, the underlying colonial ideals about science and society persist. Engaging in ongoing and evolving discussions about stories that we tell about science, including those who we uphold on pedestals, those who we have named halls, exhibits and entire buildings after—the roles that coloniality has played and continues to play in science—is necessary if we are going to change science education for teachers and students alike.

Historiography

In my practice as a museum educator in science, I used the museum's resources to support science teaching and learning, and I rarely accessed cultural artifacts. However, the cultural and historical artifacts—like the mural, statue, and interpretations of the ethnographic collections—are a part of the Museum's structure and, therefore, have implications for any teaching and learning in the Museum's space. How people perceive the Museum and these objects agency in their hidden and unpacked histories and, therefore, the ability to constrain the science teaching and learning in the spaces where they are present. As an example, if a person of African descent entered the museum, noticed the statue, and had a similar visceral response to me, it could affect how they approached the scientific objects in the museum. As a racialized space, it can be perceived as unsafe mistrusted, and may even generate anxiety about encountering additional displays and objects that have racist connotations. Furthermore, these racist and subjectively biased objects in a scientifically authoritative institution such as the Museum reinforce the ideology of Western superiority—both in science and culture—shutting the viewer down from the opportunity for expanded agency in science as "the knowledge Western science produced became the benchmark by which the productions of non-Western civilizations are measured" (Kincheloe, 2001, p. 475). I have heard of college students training to be explainers in one of the Museum's summer internship programs reading some of the text in the Hall of

African Peoples and becoming highly offended and angry at the stereotyped interpretations of Africans and African cultures. Although they continued in the program and became exemplar explainers during the summer, I have little doubt that this experience influenced how they viewed themselves within the Museum.

As an educator, I have had to deal with my own feelings about the legacy of racism both in my personal experiences and how those experiences are replicated in the historical display of people of African descent and their artifacts—I have had to keep a constant vigil against internalizing racism (Brock, 2005). For me to understand and effectively utilize the Museum as a teaching and learning resource for all students, it is important for me to learn and understand the historical constitution of the Museum as presented in the objects and the halls. As a researcher in informal science education with a critical, decolonizing stance, I have had to reconcile my feelings about race with the history of the institution and conceive new ways of using those very objects as teachable moments to help myself and other people move beyond being defined by others—"When you define your existence based on the ideals of others, you give them power" (Brock, 2005, p. 39). The oscillation between the Museum's past and present is a reflective practice of historiography, a method to "delineate the larger constructs which inform the ways [a researcher] makes sense of the past, present, and future (Villaverde, Heylar & Kincheloe, 2006, p. 316).

Through historiography, using the power of my stories in relation to the objects in the Museum, as well as the collective endeavor of the Museum to present scientific and cultural knowledge to the public, I have been able to change my stance in relation to accessing and using the Museum's and hopefully influence students and teachers wishing to do the same. A critical examination of the historical context of the Museum can be both informative and transformative. Informative in that it helps us to understand how contemporary situations came to exist in a way that informs our actions and transformative in that an understanding of these historical structures helps to develop a critical consciousness that is liberating. As hooks (1994) explains, combining the analytical and experiential is a richer way of knowing. In speaking of the critical pedagogies of liberation, by tapping into our personal experiences, we can look at/analyze the societal structures that seem to continue the domination politics in such a way that gives us a "purpose and meaning to struggle" (hooks, 1994, p. 89). It can allow us to examine the social, political, and economic conditions that are at odds with the individual's will to freedom

(Villanueva, 1993). If we think about freedom as being joyful learners unencumbered by historical oppressions, we must unpack the historical ideology of the museum and how it shaped the education programs, including the critical issues that must be addressed as we move forward with critical, antiracist and decolonial stances in science teaching and learning and science-rich cultural institutions.

Culture and Learning in Museums

Museums are often described in the context of the objects they contain and their physical structure—the visible and tangible resources as well as the buildings that contain the objects, as these structures are often of historical and/or architectural significance. However, there is also the invisible structure (schema)—the underlying ideologies that have shaped the practices of the museum, including the teaching and learning practices.

The culture that gets enacted in the museum is responsive to these visible structures and invisible schema. While the visible structures of a museum are obvious in the presence of the building and the objects that it contains, it is the invisible schema that underlies the displays of objects, the historical development of exhibits and the discourse around exhibits and education that need to be made visible. This schema contributes to the affordances available for learning in the museum, including the teaching and learning that can be enabled or constrained based on one's perceptions of the museum. While this chapter is not meant to be an exhaustive history of the Museum, I will use education as a space of analysis in order to bring to the forefront the past or history that simultaneously is present, as the history of the Museum is always present—both in visible structures, such as the old dioramas and schema, the "spirit of the institution" both of which are shaped by particular historical ideologies about nature, human categorizations and the role of the museum in society.

The Museum is an ideological space—a space that represents the ideas and interpretations from a culturally specific point of view of Western civilization, conserving the position of the European Man in his position at the apex of humans. This is implicit in an early statement about the role of the Museum—early Museum leaders saw the Museum as having the important social conviction of bringing nature to those who were not able to get out and experience it for themselves, "very few people, even among those who have the means to travel, really see nature in the sense of understanding it, and to the millions within the cities, nature is practically unknown, so we [museum scientists and exhibit developers] are *interpreters*" (emphasis mine) (Osborn, 1927, p. 269). It is curious that this contradicts another statement by Osborn (1927), "the peculiar teaching quality of a museum is that it teaches in the way nature teaches, by speaking to the mind direct and not through the medium of another mind" (p. 281). Osborn omits the role that interpretation plays in "speaking directly to the mind" as the Museum is not nature, but rather a curated display of nature. It is critical to be reminded that one of the early leaders who played a central role in curating the museum was also a eugenicist and grounded in white supremacy. He viewed the Museum as a way to display the Western, patriarchal dominance of the globe, "the institutionalization of the modern museum coincides with bourgeois disciplining and the intensification of nation-building, as well as capitalist liberalism" (Bayer, Kazeen-Kaminski & Sternfeld, 2018). Museums and other cultural institutions at the time were in the service of national identity and the idea of nationhood that had eugenicist underpinnings that excluded racial, ethnic, and socioeconomic "others". Viewing the history of the Museum and its relationship with education and schools through these lenses allows us to see how these ideologies played out in the goals and visions of museum education and how the founders of the museum saw themselves vis-à-vis diverse audiences.

Because of the position of AMNH, the underlying notion of science as being objective, the study of nature as a science, and therefore any interpretation of nature is unbiased, will also influence the perceived "authority" of the Museum. Interpretation is culturally situated—a selective activity—as it is the interpreter who decides what information gets conveyed and the interpreter as a cultural being is not free from an ideological bias. As a public entity, the Museum could be a space to begin the discussion about this ideology, and as it has done in recent decades with culture, it can bring to the forefront the idea of having different interpretations and explanations of scientific data and phenomena.

Practices, Beliefs, and Values of Science and Science Education in the Museum

The Enlightenment saw exponential growth in modern Western scientific knowledge, and correspondingly, new institutions emerged. Many museums were founded during this era as places to display collections by members of the elite class (Melber & Abraham, 2002). Charles Wilson Peale, a portrait artist, naturalist, and collector, opened the first museum in America in Philadelphia in 1784. Peale's museum was a "secular temple" where the "most perfect order in the works of a great Creator..." were displayed. Arranging objects—stuffed animals posed against painted backgrounds, wax figures dressed in traditional clothing—Peale wanted visitors to leave his museum "happily amused and certainly instructed." Describing the museum as a "temple" positions Peale's Museum as a supreme authority, with Peale a prophet of sorts who had divine power over the objects in his temple through collecting (conquering), classifying and manipulation through display. The objects were arranged and ordered through his gaze, and this, therefore, granted Peale, as a wealthy man of European descent, power over nature.

Subsequent American public museums were funded by philanthropists, such as Field, Bishop, Peabody, Carnegie, and Smithson, several of them including objects from their personal collections in their museums (Melber & Abraham, 2002). This practice of collecting nature was also a practice of accumulating power. It was the European gentry who had the capital and access to collecting voyages. They also had power through display, being able to shape the visual narratives of these objects for the view of the general public. John Smithson's collection provided the foundation for the Smithsonian institutions, his will bequeathed that his "worldly goods be used to 'found at Washington [District of Columbia] an establishment for the increase and diffusion of knowledge among men.'" (Melber & Abraham, 2002, p. 46). These museums helped to establish a colonial authoring of nature that positioned the natural history of Earth all in subordination to Man.

The Tradition of Collecting and Classifying

The early museums evolved from the cabinets of curiosities that were often objects from the private collections of the elite class. Peale, as described above was one such collector. As tools of empirical research, possessions of a diversity of objects raised the social prestige of the owner (Müsch, Rust & Willmann, 2001). These objects—both science and cultural artifacts—were collected from travels and expeditions around the globe, most of the objects originating from European colonies around the globe. The aim of early American museums was to provide audiences with an overall picture of the natural world, an objective truth about nature and human existence. However, in using a particular classifactory structure, they formed heterotopias—combinations of different places as though they were one (Kahn, 1995) and defied the lived logics of the relationality of the living Earth.

Collecting, classifying, and displaying are the hallmarks of visitors' experiences in museums. However, these practices have historical ties with the underlying and unexamined coloniality, "The critical features of taxidermy… are the hunt, the seizing of a body, which is then followed by the carving and rearranging of the now hollow vessel that is re-packaged to be exhibited as an emblem of superiority and valor" (Choi, 2021, p. 3), life on Earth, including Othered humans are reconfigured as "commodities of white supremacism" (p. 2). Thus, acts of naming, classifying, and displaying gave museums scientific authority, "all museums are exercises in classification and it is precisely from their position as 'classifying houses' that museums become institutions of knowledge and technologies of power" (Kahn, 1995, p. 324). As classification is an *interpretation* of characteristics and the creation of categories based on these interpretations, these practices reflect a specific ideology of science that is based on colonial logics of individualism (describing organisms separate from their environment and relations therein), ownership (naming gives ownership over the "discovery" of a "new" species), and exploitation (how can the benefits of a species be extracted for human use and profit). This is also the case for the science of anthropology and the Museum, as classificatory logics (Wynter, 2003) were extended to human "races" and cultures, placing them hierarchically with European man at the top. It is important to keep in mind that the history of the museum's education activity is important as the museum presents a culturally specific way of seeing the world, which often conflicts with the cultures and worldviews that students and teachers bring into the museum.

Western Science Ideology and the Museum

Western Modern Colonial Science is a culturally produced phenomena where the struggle is for the control of knowledge. It is by historical accident that the leading practitioners of science are located in the "West"; therefore, the science that is produced will be ideologically slanted to support the vested interests of Western European dominance (Kincheloe, 2001). Science is "a force of domination, not because of its intrinsic truthfulness but because of the social authority (power) that it brings with it" (Kincheloe, 2001, p. 476); a "modern folklore" conscripted into the service of maintaining hegemonic power (Villanueva, 1993). Defined as ideological domination by consent, this hegemonic view of science, is represented in museums and other institutions as markers of what exists in "civil society."

Museums help to mold much of what we understand about science, "science museums in the United States have remained attached to merely presenting materials as wonders of nature or as technological feats" (Vackimes, 2003, p. 8). This presentation of materials represents a particular cultural stance as classification, interpretations, arrangements, and visual text are all culturally situated and in the service of advancing the ideals of Western modernity.

Like other settler-colonial institutions, the American Museum of Natural History (AMNH) was founded on the ideologies and power of the dominant culture at the turn of the 20th century. In preserving the descriptive statement of Man, the museum displayed the perfect order of nature with the European, affluent man at the vertex. It is an artifact of the social construction of science produced in a particular culture in a specific historical era." (Kincheloe, 2002, p. 472)

The Foundations of Education in the Museum

The American Museum of Natural History is one of the oldest of its kind in the United States. As with other American museums, the American Museum of Natural History was founded with the "notion of public education clearly in mind (American Association of Museums, 1984, p. 55 in Melber & Abraham, 2002). As articulated in the mission statement, the American Museum of Natural History aimed to make the education of the public central to its operations.

> The American Museum of Natural History founded April 6th, 1869, for the purpose of establishing in said city a museum library of natural history; of encouraging and developing the study of natural science; of advancing the general knowledge of kindred subjects, and to that end of furnishing popular instruction. (AMNH, 1930)

The founders saw the importance of conveying to the public the great work and discoveries of the Museum's scientists and collecting expeditions. One of the founders who continued to make his mark on the development of the Museum's education programs was Albert S. Bickmore. When the Museum opened its doors in 1878, Morris Ketchum Jesup was the President, and Bickmore served as the Superintendent. Jesup saw the value of the Museum's content as going beyond a monetary value, and it was his belief in the ideals of science and his desire that "they [the museum's content] should be brought within the comprehension of all classes of people" (Osborne, 1911, p. 29).

> The value of what you have already accumulated in your halls rises to a large figure commercially, but it is a difficult task to estimate the money value of what belongs to science and scientific institutions. To their values must be added their ameliorating power, their educational force, and the scope they afford the higher faculties of man to apprehend the wonderful phenomena of Nature, and to master and utilize her great forces. (From an 1884 administrative report in Osborne 1911)

This statement clearly articulates the ideologies about nature and science of the time. Violence, extraction, and colonization are evoked by the words "apprehend," "master", and "utilize." Furthermore, the pronoun "her" is attributed to nature; this reveals the cultural-historical entanglement of the femme and nature (Braidotti, Charkiewicz, Hausler & Wieringa, 1994). Gendered as a female, nature loses any agency and becomes in servitude to Man.

The founders of the museum sought to extend the value that they saw in nature and as represented in the halls of the Museum to the city's schools. To these founders, all white men from the upper class, the museum collections had ameliorating power, the ability to improve the lives of all people who encountered these objects, artifacts and displays. By putting the museum's collections within reach of the city's children and schools, the museum was positioned to improve the lot of many students, especially those who were labeled as economically and, as an extension, intellectually impoverished. This compelled Jesup and the Board of Trustees to establish the Department of Public Instruction, which was officially constituted in 1884. This department would encompass "all those features of the Museum which are instrumental in articulating the work of the Museum with the public at large and especially with the educational system of the city" (Osborne, 1911, p. 116). Through the creation of this Department of Education, the vital link between New York City public schools and the Museum was established.

In 1880, the New York State Department of Education enthusiastically approved the Museum's proposal to offer a series of lectures to Primary School teachers and principals. Thus, in January 1881, Bickmore began a series of lectures in zoology and natural history, establishing him as the Museum's first public educator. In 1884, grant money from the State Department of Public Instruction was appropriated "to establish and maintain a course of free lectures to the teachers from the common schools of New York City and to the teachers of the common and normal schools throughout the State, who wish to avail themselves of this training" (Quotes from the grant proposal in Osborne 1911). Professor Bickmore was appointed as the Museum's first Curator-in-Chief. The school year of 1884 was greeted with ten lectures on physiology, zoology, and botany in a course of study presented by Professor Bickmore.

The Bickmore Slides—Providing a Resource for Teachers and Schools

Serving as the first curator of the Department of Education, Bickmore is credited with developing a series of lantern slides and lectures on natural history for public school teachers, establishing the idea of the museum as a resource for the city's teachers and schools. Inspired by his youth growing up in the woods of Maine, Bickmore felt that what he called "the visual method" of education was a vital way of allowing students and teachers who don't have access to nature to be able to see and make observations and connect with the natural world. This is evidence of early notions of "nature deficit disorder" (Louve, 2005) described to be unevenly experienced by urban children, who are then positioned as lacking in other ways that would impede their advancement in society. In 1895, New York State issued an act to provide "The Visual Instruction Method" to common schools within the State, and accordingly, access to Bickmore's lectures became far-reaching. Bickmore reproduced "lantern slides" from the best photographs taken on the Museum's famed expeditions for the lectures and distribution. Bickmore himself traveled to "remote lands" of the world to gather information and take photographs. He also collected negatives from world travelers, "there was not a traveler of note who came to New York, whom he did not seek out and ask for negatives" (Sherwood, 1927, p. 317). Professor Bickmore's slides, some of which were "beautifully colored," served as the basis on which the Museum's lantern slide collection was developed.

The lantern slides served as an important piece in the Museum's relationship with New York City Public Schools. Beginning in 1892, the Museum offered lantern slide lectures to teachers; they were so well-attended that a theater was built for this purpose, it serves as the IMAX theater today. In 1908 the Museum started to advertise motion pictures to accompany the lantern slide lectures. However, the then president and eugenicist Osborn felt that the motion pictures were "frivolous" and took away from the educational nature of the static lantern slides (Wray, 2010).

In 1915, 20,000 slides were made available for school use; teachers were able to pick and borrow them from the catalogue to complement lessons in the classroom. Teachers and students were also able to visit the Museum's library to study the slides on site. As the popularity of the lantern slides grew, special themed collections were formed:

> Slides were also available in boxed series with titles such as ... "Our Atlantic Coast Fisheries," "Pond Insects." Boxes of slides were shipped throughout New York State, and trucks delivered slides, taxidermy wildlife specimens, and miniature dioramas to New York City schools; a motorcycle with a sidecar was available for quick delivery. (Wray, 2010, p. 19)

Bickmore served as the curator of the Department of Education until 1904. During his leadership, the Department's service focused on instruction for teachers. This instruction included evening lectures on natural history, geography, and industry supplemented by the Bickmore slides. He officially retired in 1906 due to failing health, but his special collection of lantern slides known as the "Bickmore Slides" continued in circulation to the city's schools.

From the inception of American museums to early efforts to supplement school-based studies with museum resources, natural history museums played a central role in shaping public ideologies around nature and science. The natural world was positioned as something that was removed from the lives of urban students; little attention was paid to the nature that surrounded them in the city. For Osborne, the city was dirty; therefore, neither had value in education nor was integrated into the "spirit of the institution." Educational initiatives were conceived on this foundation that devalued the inhabitants of the city, especially the poor, immigrant, racialized groups, and the city itself. In essence, the museum was an artifact of the descriptive statement of Man and all structures and practices therein served to conserve this—from the mission of the institution to the on-the-ground programs to schools and students.

Connecting the Museum to School Curriculum

During the time of the establishment of the Museum's education services, the New York City public schools were under the leadership of Superintendent William H. Maxwell. Maxwell who is credited with establishing a broader and more uniform curriculum in the public schools. He sought to give the city's children as many educational opportunities as possible and was against reformers who compared the education of children to that of manufacturing industrial goods (Ravich, 2001). He expanded the role of the public school by adding afterschool programs and services for kids with disabilities. Maxwell also served as the chairman of the Committee of Fifteen, where William T. Harris was the head of the subcommittee to deal with the correlation of studies. The committee was established with the goal of creating a common curriculum for New York City public schools. This Committee championed the cause for a humanist curriculum "constructed around the finest resources of Western Civilization" (Kliebard, 1987, p. 17), with the foundation laid by people of noble hereditary character, according to the eugenicist Osborne.

While Maxwell had an expanded view of the purposes of education, the educational context remained structured around supremacist views of European civilization, including the museum's resources. The city also increased its financing of the Museum's operations (after the State withdrew financial support in 1904), and the Museum felt that it was "proper…to give its attention, first to the needs of the City's schools…" (Sherwood, 1927, p.317), which was then endorsed by Maxwell's administration. At the request of the New York City Teacher's Association, the Museum commenced a series of lectures to school children in supplementation of classroom work in geography, history, and natural science.

A nature study curriculum for elementary school was established as part of the city's curriculum reforms. The Museum was able to supplement this curriculum with the oldest and, at the time, the most extensive aspect of its service to schools—the circulation of its nature study collection. Wooden cases containing representative specimens of various animals, such as mammals, birds, and insects, and samples of minerals and woods, as well as public health charts and exhibits, were made available on loan to schools. They were designed so that the specimen could easily be removed and handled by students. Also available were Museum-developed mini-dioramas called "habitat group types." One example, "Birds that are Our Friends," presents a group of native birds, including a Screech Owl with a mouse in its beak displayed against a painting of the natural environment and types of trees and plants that would be found in the habitat. The nature study collections were accompanied by literature describing the animals and their relationships to each other and to humans, as well as a bibliography of popular books on the subject. These teaching tools were available, free of charge, on loan to schools.

The Sherwood Era: Expanding the Museum's Service to Schools

In 1906, George H. Sherwood began his service as Curator-in-Chief of the Department of Education. Described as a practical teacher who believed that the "training of children is the most important vocation in the world" (Sherwood, 1927, p. 320) under Sherwood's leadership, the Museum expanded its service to the schools. Sherwood believed that "it is the function of the Department of Public Education to digest this material and to present such portions of it as will be useful to teachers and pupils." Sherwood (1927) distinguished two main branches of the Museum's service to schools. The extramural branch involves services that occur in places outside of the Museum plant and included the lantern slides service and circulating nature study collections to branch libraries and branch schools, both began in 1904 and 1914, respectively. Of the lantern slide service, Sherwood believed that "the use of the slide in the classroom and assembly simplifies the teacher's task and enables the pupil to absorb information more quickly and permanently" (Sherwood, 1927, p. 327). Sets of slides on curriculum topics were developed and accompanied by lecture manuscripts, enabling teachers with limited preparation time to use the slides with minimum effort. The distribution of films and lectures to the schools was also included in the extramural services. Special exhibits were made available for loan to public libraries. The primary purpose of these exhibits was "to stimulate children to read good books." (Sherwood, 1927, p. 329). These exhibits included specimens such as animals, artifacts and industrial models that could be used to illustrate books on travel, geography, nature study and a host of other subjects. The intramural branches were services that took place within the Museum and included lectures, a service for blind students, and instructors in the exhibition halls. While the desire to make resources applicable to schools and students is necessary for museums, this was still in an era where banking models of education dominated, especially for the populations of students in the city's schools. As a result, there was little agency offered to teachers in how they could or would adapt resources to their classrooms and even less for the students who would learn with and from these resources.

Sherwood found the nature study collections most valuable to teachers and students. He recounted that teacher found them useful in teaching facts about nature and doing language work, particularly in teaching English to immigrant students. He believed that the greater service was "giving city children a glimpse of the great outdoors" (Sherwood, 1924, p. 272). The city's public school children were viewed as impoverished, not only in terms of capital but also in terms of exposure to different experiences, including the "great outdoors." The narrative of the "great outdoors" was central to the American lexicon of wilderness experience as minimally touched by humans. For urban students, this meant a science curriculum disconnected from the world that they experienced around them.

According to Sherwood, if children rarely left the city, they had no references to the great outdoors and offered this example:

> The class was reading a poem dealing with the "signs of spring," daffodils, frogs, etc. The children did not comprehend the meaning of the phrase [signs of spring]. Finally, the teacher asked how do we know that spring is here? Johnny was the only one who raised his hand. "Well, Johnny, how do you know that Spring was here?" "Because I saw them hanging the swinging doors on the saloons." Certainly, the nature study collections from the Museum helped to give Johnny a new conception of spring. (Sherwood, 1924, p. 272)

While it is not known if this were an actual or fictional event, clearly Sherwood saw little value in the students' urban lived experiences and lived understandings of the change of seasons. Saloons in NYC at the time were associated with immigrants and criminality, and by using this example in contrast to flowers and frogs, Sherwood established the superiority of the "new concept of spring." Sherwood emphasized the perceived deficits in Johnny's life that would be greatly improved by the nature study collections and saw the "service to schools" as a means to doing so.

Sherwood (1927) described circulating nature collections as having the ability to "awaken the spirit of research" and encourage students and their families to visit the Museum and go back to the library for further reading. This program formed the basis of cooperation between the libraries and local schools, thus increasing the accessibility of the Museum's collections to local communities around the city. As a part of its extramural services, the Museum also provided lectures to schools, but due to limited staff, these mainly took place in centrally located schools where kids from surrounding schools could visit and have the benefit of a quality Museum lecture, supplemented by slides and specimens from the education collection, without the expenditure of carfare. While Sherwood was concerned with students and their families having access, it was assumed that the objects alone would motivate students and their families to visit museums and do library research, confirming a myopic view of what was important to the city's families at that time or imposing a view of what *should* be important on others.

In 1926, over 170,000 students and teachers attended lectures at the American Museum of Natural History. As a part of its intramural activities, these lectures provided a valuable resource "designed to supplement the work of the classroom teacher, not to replace it" as Sherwood mentioned on numerous occasions (e.g., Sherwood, 1923). Using the New York City public school curriculum as a guideline, these lectures were designed for elementary students to learn about history, geography, and natural science. The lectures illustrated with the lantern slides and corresponded, whenever possible, with exhibits at the Museum. Upon questioning the value of the lectures, Sherwood found through testimonies that teachers found great value in them. He noted that one teacher even asked for the unused lecture programs, which she distributed to her students to take home. These lectures also sparked interest in the parents of the subject matter, one mother expressed that the Museum lectures so inspired her children that they could not stop talking about them, and consequently, "they all gathered around the table at home to read the story books based on the Museum lectures" (Sherwood, 1927, p. 331). For the high school students, special lectures were given during the week of the state mandated exams on biological topics after which the students were sent to the exhibition halls with questionnaires for further study. The Museum also provided a service of instruction and guidance in the halls. If a school booked a trip a few days in advance, they would benefit from an instructor teaching their class in the halls. These instructors were also present in the halls during opening hours to teach visitors in front of the exhibits.

Through a special endowment, the Museum dedicated a special branch of its work for the education of blind students in the New York City public schools. Under the guidance of a special Supervisor for the Blind, the Museum's visual education program was adapted to include recordings of the lectures and tactile interactions with the nature study collections. "The results from this work [are] gratifying," Sherwood (1924) wrote, "often they are read in the children's happy faces." One teacher wrote of the impact the trip had on her students:

> Children of little experience in life and meager opportunity for general information speak with ease and familiarity of animals, birds, people, and customs. You can scarcely realize to what extent you are broadening their horizon. (Sherwood, 1924, p. 272)

We see that not only museum leaders but also teachers held deficit views of urban students. The assumption is that their daily experiences were not rich and varied.

My father was born in 1924 in New York City to Caribbean immigrant parents around the time when the Museum ramped up services to schools. His family first lived on the Lower East Side before moving to Harlem, northeast of AMNH. He used to tell us about class visits to the Museum when he was in elementary school and how he found the dinosaur fossils fascinating. While the large, extinct animals sparked his curiosity, he was unable to pursue research at home. Being a lower-income family during the Depression, there were other priorities, like buying food and paying rent. However, my father regaled to us many stories of growing up in New York City in the shadow of AMNH. Although his family neither had the financial resources nor the dominant cultural resources of the likes of Sherwood and the teacher Sherwood cited, my Dad's experiences were full of adventure. He swam in the East River and climbed trees in Morningside Park. He had rich and varied interactions in urban nature. These interactions and experiences were disconnected from his schooling and visits to the museum. According to the Museum, nature was not something one encountered on the streets of Harlem.

In 1909, the City's Board of Estimate and Apportionment designated funds for the construction of the School Service Building. This building would house the lantern slides, collections, and film distribution services, as well as provide spaces for classroom work and other educational activities. This building, according to Osborne, would provide the means to accomplish what he describes as the chief public mission of the Museum, "to bring the wonders and beauties and truths of Nature from every land and every sea, to exert their broadening and uplifting influence" (Osborne, 1923, p. 3).

Creative Education

Osborne was appointed as President of the American Museum of Natural History in 1908. He was an anatomy and physiology professor from Princeton University and based his theory of "creative education" on his experiences teaching there. As president of the museum, I found that this theory of "creative education" facilitated changes in the presentation of Museum exhibits that, in turn, influenced the Museum education pedagogy. His book *Creative Education in School, College, University, and Museum* advanced his ideas on education that situates the learner in an active role in their education with the role of the teacher to create the context for increasing student agency:

The factors of education (emphasis his) are the processes of storage of these forces by the cooperation of teacher and student, the former with his constantly diminishing, and the latter with his constantly increasing. (Osborne, 1927, p. 311)

He published this book when Dewey's ideas about democracy and education were already in circulation, and while there are similarities in centering student agency, Dewey espoused education for the advancement of a fair and equitable society. Dewey also advocated for object/inquiry-based learning, an approach that remains central to museum-based learning today (Adams & McCullough, 2021). Osborne claimed that the basis of education, the "creative and productive," is built on seven principles: truth, beauty, learning, observation, reason, expression, and production. It is interesting that he mentions, "The principle of seven cardinal elements of education is my own…it is the product of fifty years of experiment and observation as a teacher, not by reading what other people have written about education" (Osborne, 1927, p. 311). With Dewey's theories about education being popular at the time, and given Osborne's interest in education, it would be implausible to think that he did not come across Dewey's work. However, given Osborne's stance on the superiority of the white race, he most likely aimed to distance himself from Dewey, a proponent of progressive education and social progress. It is also interesting to note that Osborne developed these theories based on his work at Princeton University, a school for the white, mostly male, elite-noble class. While he often discussed the ability of nature to inspire students, I did not encounter literature where he suggested the same degree of agency in learning to urban students and the teachers who taught them. In other words, learning both in content and approaches was designed to conserve the descriptive statement that places Osborne's race and gender at the pinnacle of human. The elite classes were to have education approaches that allowed them to have increasing agency over their learning, whereas the *cloaca gentium* would learn in ways that were more didactic and transmissive and emphasizing obedience to "authority".

The museum was a great silent teacher to Osborne, "…every specimen, every exhibition, every well-arranged hall speaks for itself" (Osborne, 1927b, p. 240). As president of AMNH, he played a critical role in the museum's seminal design. He was greatly influenced by his professor Louis Agassiz, who was also an eugenicist. Osborne often quoted Agassiz's motto, "study nature not books." Agassiz believed that students should first make direct observations before consulting a teacher or textbook. Osborne transferred this philosophy to the Museum, where he paid great attention to the arrangement of the exhibits and the halls. He wanted the Museum to be a place to "bring a vision of the world to those who otherwise can never see it" (Osborne, 1927b, p. 244). Undoubtedly, his ideologies around white supremacy were central in creating this vision. The halls, along with the objects, taxidermy, paintings, and photography they entail, served as "vessel[s] for colonial ideologies such as Manifest Destiny that frame the dominant white culture as civilized and that of the ethnic Other as wild and inferior" they are "manifestations of a colonial history and power, are concerned with external expressions of 'realism and idealization' shaped by Western hegemony" (Choi, 2021, p. 870).

Today, deficit narratives still undergird many STEM education-related initiatives targeted at historically minoritized learners. This was no different in the early years of the education programming at AMNH; the focus was on connecting "children of little experience and meager opportunity for general information" (Sherwood, 1924) with nature. When I visited the museum archives to look for photos of children learning in the Museum, I found that most of them depicted poor and immigrant students utilizing the Museum's nature study programs. The idea was that these students would be uplifted through having the experience of seeing and touching these natural objects. There was little notion of them bringing their personal experiences to bear in their interactions with objects and exhibits—according to educators at the time, these children did not have any valuable personal experiences to share (Figure 2.2).

Towards Decolonizing the Museum

In the nascent 21st century, the Museum has been taking action to address some of the problematic histories. Tomanowos, or the Willamette Meteorite, is representative of the Sky People and a sacred being to the Clackamas. Rainwater, which collected its nooks and crannies, was healing and empowering; existing songs and dances describe the relationship between the Clackamas and the meteorite. In the 1850s, the Clackamas, along with other tribes and bands, were forcefully removed from their ancestral homelands in what is now called the Willamette Valley to the Grande Ronde Reservation. This left Tomanowas alone and on stolen land that became under the ownership of the Oregon Iron and Steel Company. Through a series of thefts and exchanges of capital, Tomanowas found itself at the Museum on display.

Figure 2.2. Children studying nature in the Museum. © The American Museum of Natural History

Tomanowas was not forgotten by the Grand Ronde. In the 1990s, the tribe, Oregon residents, and politicians fought to have Tomanowas returned to the Grand Ronde. Actions and lawsuits evoking the *Native American Graves and Repatriation Act* were leveraged, and the Museum filed a countersuit that contested the Grand Ronde's claim. With an out-of-court settlement, both sides agreed that Tomanowas would remain in New York, with the Grand Ronde people having access to private annual ceremonies. Each year two Grand Ronde youth are selected to participate in a museum internship program. The text accompanying Tomanowas describes both the scientific and sacred significance of this object. To scientists, it is a 15.5-ton iron object from space. It is one of the largest meteorites found in the United States and provides information about the formation and composition of the universe. To the Grand Ronde, it is a sacred, powerful, and living being.

Desettling Objects with Teachers

One of the workshops I designed and facilitated for science teachers was about using the cultural halls to teach science. First, by defining science as any systematic attempt to produce knowledge about the natural world (Semali & Kincheloe, 1999), I explained that based on this definition, *any* system of gathering knowledge is considered science, including indigenous systems. I then presented the ideology of modern, Western-based science, including the Cartesian dualism of separating the knower from the known (Kincheloe, 2001) and establishing norms, "positivism presupposes that reality is singular and objective" (Reagan, 2004, p. 17). This framing begins the discussion of questioning the interpretation of cultural artifacts and resituating them as objects of advanced innovation and technology for the time and context when they were developed. To observe and reflect on the objects, I posed the question, "What science knowledge scientific knowledge was needed for the creation of this artifact?" Through our encounters with the artifacts and discussions, we realize that Indigenous people had to know and understand a great deal about the natural world in order to create the artifacts on display. People of East Africa had to know about animal behavior to create appropriate hunting and fishing tools. Indonesian people had to know what we call chemistry in order to use plants to create dyes for fabrics and body paints. Pacific Islanders had to know physics to develop sails and navigation tools and material science to morph the coconut into its many forms and uses. This indigenous knowledge, as Semali and Kincheloe (1999) explain, is "the dynamic way in which the residents of an area have come to understand themselves in relationship to their natural environment and how they organize folk knowledge of flora and fauna, cultural beliefs, and history to enhance their lives." (p. 3). This leads to a discussion of the relationship between knowledge and power in relation to both science and education more broadly,

> The power struggle involves who is allowed to proclaim truth and to establish the procedures by which truth is to be established; it also involves who holds the power to determine what knowledge is of most worth and should be included in academic curricula. In this context, the notions of indigenous knowledge as "subjugated knowledge" emerge to describe its marginalized relationship to Western epistemological and curricular power. (Semali & Kincheloe, 1999, p. 31)

Teachers who participate in this workshop often find seeing the science in the cultural artifacts liberating in that it provides an interesting way of integrating cultural studies with science. It opens up the possibility of dialogue about what science is and who decides what gets to count as legitimate knowledge. It also allows teachers and students to examine their own indigenous, cultural and community practices and beliefs and compare them to the science that is taught in school. However, this is an enrichment workshop, one that is neither central to teaching and learning in the Museum nor a part of the classroom science curriculum.

During the past 20 years, museums have been considering new scholarship in order to interpret cultures from more egalitarian and more authentic perspectives (Levin, 2002). Museums have moved away from ideological science as the only way of interpreting objects. In an exhibit about Haitian Voudou, the Haitian community played a central role alongside the Museum's anthropologists in curating the exhibit—deciding how the objects would be displayed and described. However, this has not been the case with science. During my time in the museum, ventures into the cultural halls were not a part of the usual science education agenda; this was reserved for teachers who were teaching social studies, literacy, and the arts, reinforcing the narrow association of science with Western, Eurocentric culture. Furthermore, not having a people hall dedicated to Europeans maintains the hegemony of "the west" as the knowledge producer, as represented in the building within which other races and cultures are positioned as exotic, primitive, and subordinate to the authors of their descriptions. This also reinforced the idea of science as a singular objective truth rather than diverse ways of knowing and understanding how the natural world works.

Making Historicity Visible

The global witnessing of George Floyd's murder in 2020 led to a new awareness within individuals and organizations about the systemic injustices experienced by Black people within and across levels of society. This has compelled many organizations, especially those involved in education, to examine their ideologies and practices with a critical lens. One of these areas for the Museum was the Theodore Roosevelt statue, which served as a symbol of settler colonialism and the corresponding dehumanization of Black and Indigenous people, along with the theft of their bodies and land. Although there had been decades of activism that targeted the removal of this statue, it was institutional self-reflection vis-à-vis a global movement of racial reconciliation that finalized its removal. Noting the disturbing racial hierarchy communicated with the statue,

> To understand the statue, we must recognize our country's enduring legacy of racial discrimination—as well as Roosevelt's troubling views on race. We must

also acknowledge the Museum's own imperfect history. Such an effort does not excuse the past, but it can create a foundation for honest, respectful, open dialogue (AMNH, n.d.).

Along with the public-facing reconciliatory actions, there is an expanding practice of addressing problematic histories and practices of the Museum through scientific convenings and the ways that learning happens in the halls and with objects.

In 2021, one hundred years after the Second International Eugenics Conference, the museum issued a statement addressing its relationship with eugenics, "the Museum welcomes the opportunity to acknowledge, confront, and apologize for its role in the eugenics movement," noting, "the pseudo-science of eugenics and the ways that it has been applied against vulnerable populations are antithetical to the values, mission and ongoing work of this Museum." The statement goes on to describe the ways that eugenics has harmed and continues to influence society and highlights some of the steps the museum is taking to address its contribution to this violence, such as ethical data collection and providing scientifically reliable information on current topics such as climate change and biodiversity loss. In concert with this statement, the museum also held an "anti-centennial" conference entitled "Dismantling Eugenics: A Convening," as described on the convening website:

> *Dismantling Eugenics* was a free, online event that reckoned with the history of eugenics; lifted up the grassroots movements working to counteract the oppressive legal and social structures that still further eugenicist ideals today; and offered space to envision and work towards an anti-eugenics future.

The event was organized in collaboration with participants from the disability justice community, ensuring accessibility, and featured artists, scholars, activists as an "interdisciplinary effort working to understand and bring awareness to the continuing legacy of eugenics, which applied [white] supremacist principles to a wide range of hard and social sciences and public policies…that continue to plague our society in the forms of racism, misogyny, ableism, homophobia, xenophobia, and other forms of social tyranny." White was not included in the original text, but I bracketed it because the eugenics movement used the standard of apex whiteness on which to build their pseudo-science theories and subsequent policies and actions. Supremacy, in this context, cannot be separated from whiteness.

In education practices, museum educators are devising ways for learners to view objects through a social and historical, as well as a scientific lens. For example, learners are prompted to learn about and reflect on the history of Tomawonas and how it got to the Museum, using the truthing language of stolen objects and stolen land and asking learners to consider culture and conflicts in relation to the objects. There is more of an effort to integrate the scientific, historical, and cultural significance of objects and halls, as well as connections to land and place, and note the information and perspectives that are left out. In this sense, the objects are afforded a degree of agency in revealing multiple aspects of their stories rather than being positioned as static, non-agentic props of science. While these actions are not yet widespread, they provide glimpses into what reconciliatory museum learning could look like, where the objects and exhibits become points of both critical and scientific interrogation. In other words, object-focused learning in museums could be facilitated to confront challenging histories while exploring science, demonstrating that science and scientific endeavors are not absent of problematic social histories and consequences.

Science museums also recognize the importance of emphasizing the nexus of science, art, and culture as a way of knowing and connecting with broader audiences. One example is the Museum of Science and Industry in Chicago. Black Creativity is an annual event that celebrates the contributions of African Americans to art, science, technology, and innovation. It started in the 1970s, and each year, it curates a series of exhibits and performances that highlight Black Creativity, including the contributions of Black architects and scientists to their respective fields. The central exhibit is a juried art show that displays the work of emerging professional Black artists from around the world. Such exhibits and learning experiences are important in identity development for Black youth and identity affirmations for Black adults. Because of the cultural authority of museums, this also centers the brilliance of Black people as agentic authors of history while also legitimizing the arts as a pathway to and a way of expanding science.

Expansivising History

Museums established in Europe and North America (and many other colonial outposts) in the 18th and 19th centuries are undoubtedly centered on whiteness-coloniality in the telling and interpretation of natural and human histories and history of science. Social movements such as Decolonize This Place and Black Lives Matter are forcing institutions to revisit and retell painful truths about their own histories in ways both reveal these histories and invite plural perspectives in the retelling of history in ways that are truthful and from diverse perspectives. While this process is neither linear nor without resistance, it is an important journey to undertake if we are to have museums and corresponding learning opportunities that are relevant and meaningful to all learners and allow for hope toward more equitable, just, and planetarily healthy futures.

References

Adams, J. D., & McCullough, S. (2021). Inquiry and learning in informal settings. In Clark A. Chinn & Ravit Golan Duncan (Eds.), *International Handbook of Inquiry and Learning* (pp. 358–370). New York: Routledge.

Adams, J. D., & Weinstein, M. (2020). Sylvia Wynter: Science studies and posthumanism as praxes of being human. *Cultural Studies↔ Critical Methodologies, 20*(3), 235–250.

The American Museum of Natural History. (2021, September). Museum Statement on Eugenics. https://www.amnh.org/about/eugenics-statement

The American Museum of Natural History. (n.d.). *Addressing the Statue.* https://www.amnh.org/exhibitions/addressing-the-statue

The American Museum of Natural History. (1930). *Preliminary Statement Sixtieth Anniversary Endowment Fund.* The American Museum of Natural History.

Bayer, N., Kazeem-Kaminski, B., & Sternfeld, N. (Eds.). (2018). *Curating as anti-racist practice.* Aalto University, School of Arts, Design and Architecture.

Braidotti, R., Charkiewicz, E., Hausler, S., & Wieringa, S. (1994). *Women, the environment and sustainable development: Towards a theoretical synthesis.* Zed books.

Brock, R. (2005). *Sista Talk: The personal and the pedagogical.* New York: Peter Lang.

Cabinet of Curiosities. (2006). In *Earmarks in Early Modern Culture.* Retrieved May 29, 2006, from http://earmarks.org/archives/2006/04/16/66.

Choi, S. (2022). "Life, Death, or Something in between: Photographic Taxidermy in 'Get out' (2017)." *Quarterly Review of Film and Video, 39*(4), 867–889.

Decolonize This Place. (2016). Decolonize this museum: Anti-Columbus Day Tour brochure and Map. https://decolonizethisplace.org/downloadable-materials.

DeSalle, R. (2021, Dec-2022, January). The Eugenics movement in retrospect. *Natural History,* https://www.naturalhistorymag.com/features/093896/the-eugenics-movement-in-retrospect

Eugenics Record Office. (1927). *Eugenics seeks to improve the natural, physical, mental, and temperamental qualities of man.* Washington, D.C.: Cold Spring Harbor Laboratory, Carnegie Institution.

Hartman, N. (2016, July). "The Passing of the Great Race" @ 100. *Public Books.* https://www.publicbooks.org/the-passing-of-the-great-race-at-100/

hooks, B. (1994). *Teaching to transgress education as the practice of freedom.* Routledge.

Kahn, M. (1995). Heterotopic dissonance in the museum representation of Pacific Island Cultures. *American Anthropologist*, New Series, 97, 324–338.

Kincheloe, J. (2001). *Getting beyond the facts: Teaching social studies/social sciences in the twenty-first century*. New York: Peter Lang.

Kliebard, H. M. (1986). *The struggle for the American Curriculum, 1893–1958*. New York: Routledge.

Levin, M. (2002). Museums and the democratic order. *The Wilson Quarterly*, 26, 52–65.

Lynn, C. (2019, March 17). Pieces of sacred meteorite—Largest in U.S. history—Returned home to Grand Ronde tribes. *Statesman Journal*. https://www.statesmanjournal.com/story/news/2019/03/07/sacred-willamette-meteorite-tomanowos-pieces-returned-grand-ronde-tribes/3056468002/

Melber, L. M., & Abraham, L. M. (2002). Science education in U.S. natural history museums: A historical perspective. *Science & Education*, 11, 45–54.

Moore, P. (2021, January). A Liberatory framework: Critical race theory can help museums commit to anti-racism and combat anti-blackness. *American Alliance of Museums*.

Müsch, I., Rust, J., & Willmann, R. (2001). *Albertus Seba's cabinet of natural curiosities*. Berlin: Taschen.

Norrgard, K. (2008). Human testing, the eugenics movement, and IRBs. *Nature Education*, 1(1): 170.

Noisecat, J. B. (2019, September 13). The environmental movement needs to reckon with its racist history. *Vice*. https://www.vice.com/en/article/bjwvn8/the-environmental-movement-needs-to-reckon-with-its-racist-history

Osborn, H. F. (1911). *The American Museum of Natural History: Its origin, its history, the growth of its departments*. New York: The Irving Press.

Osborne, H. F. (1927). *Creative education in school, college, university and museum*. New York: Charles Scribner's Sons.

Osborne, H. F. (1927). Creative education. *Natural History*, 27, 309–314.

Pogrebin, R. (2020, January 19). Roosevelt statue to be removed from museum of natural history. *The New York Times*. https://www.nytimes.com/2020/06/21/arts/design/roosevelt-statue-to-be-removed-from-museum-of-natural-history.html

Public History Project. (2021). Dismantling eugenics: A convening. https://antieugenicsproject.org/

Ravitch, D. (2001). *Left back: A century of battles over school reform*. Simon and Schuster.

Reagan, T. G. (2004). *Non-western educational traditions: Alternative approaches to educational thought and practice*. Routledge.

Semali, L., & Kincheloe, J. (1999). Introduction: What is indigenous knowledge and why should we study it? In L.

Semali & J. Kincheloe (Eds.), *What is indigenous knowledge?: Voices from the academy* (pp. 157–178). New York: Falmer Press.

Sherwood, G. H. (1927). The story of the museum's service to schools. *Natural History*, 27, 315–350.

Sherwood, G. H. (1929). The American museum in school service. reprinted from *American Childhood*. March 1929.

Treishman, R. (20, January 2022). New York City's natural history museum has removed a Theodore Roosevelt statue. *National Public Radio*. https://www.npr.org/2022/01/20/1074394869/roosevelt-statue-removed-natural-history-museum

Vackimes, S. (2003). Of science in museums. *Museum Anthropology, 26(1)*, 3–10.

Villanueva V. (1993). *Bootstraps: From an American academic of color*. National Council of Teachers of English.

Villaverde, L., Kincheloe, J., & Helyar, F. (2006). Historical research in education. In K. Tobin & J. Kincheloe (Eds.), *Doing educational research—A handbook*. (pp. 313–348). Rotterdam: Sense Publishers

Wray, S. (2010). The American Museum of Natural History's magic lantern slide collection. *The Magic Lantern Gazette, 3*(v. 22), p. 19.

Wynter, S. (2003). Unsettling the coloniality of being/power/truth/freedom: Towards the human, after man, its overrepresentation—An argument. *CR: The New Centennial Review, 3*(3), 257–337.

· 3 ·

TEACHER AGENCY AND IDENTITY: CREATING AFFORDANCES IN THE EXPANSIVISING SPACE

I met Violet and her classes in the Hall of Planet Earth on one of her field trips to the Museum. She brought three of her classes at once—about 80 students—because her administration would only allow her to take two field trips for the year. Each of her students had a brightly colored folder containing activities to complete while at the Museum. Her focus was Earth Science, and her goal for the trip was to have students generate research questions for their Exit Projects. She divided her classes into three groups and rotated them between three Earth Science-oriented halls. She asked me if I could tag along with the math teacher's group in case the students had any questions that she could not answer. Before I left Violet with my group, she gathered her students in a central space in the hall to let them know what was expected of them.

"You are to visit the halls and complete the activities for each hall. You are working in your groups, so you could work together and share your observations. We went over the activities in the class, so if you have any questions, you could ask Ms. Adams or me." As the students gathered in their working groups within their classes, I took the opportunity to shuffle through the folder so that I would know what the students were expected to do. There were three stapled documents, each labeled with the name of the hall the students were to explore and contained a variety of closed and open-ended questions with plenty of room for writing and drawing observations. I recognized some of the questions, as they were the same ones given to teachers during the Urban Advantage workshops.

Urban Advantage: Leveraging the City's Science-Rich Resources

Violet was a teacher in the Urban Advantage initiative (see Chapter One). She started in the first year of the program and was very excited about the opportunities it offered her as a teacher and for her students. Violet was at the beginning of forming the expansivising space between informal museum learning and the science classroom. Expansivising describes the limitless possibilities for configurations of learning with different resources in relations. As Violet learns about the affordances in ISE settings, she will use her lived professional-practical knowledge about the affordances in her classroom—resources, policies, curriculum, and, importantly, her students, to help create new affordances in the expansivising space. Her increasingly central involvement in developing these new affordances shaped her agency and identity as a teacher. This also expanded opportunities for her students' learning.

Urban Advantage was structured around the eighth grade Exit Projects. These are long-term science investigations by students required for graduation from middle school. The projects could be based on the three overarching content areas learned during middle school: Earth science, Life Science or Physical Science. As such, each of the seven partnering institutions[1] chose a content area of focus based on the exhibits, behind-the-scenes research and activity and resources available at their site. The AMNH focused on Earth Science. The Bronx Zoo, Staten Island Zoo, Brooklyn Botanic Garden, and New York Botanical Garden emphasized Life Science, and the New York Hall of Science facilitated learning in the physical sciences. The Urban Advantage teachers chose a primary institution and two options of secondary institutions. At the primary institution, teachers would complete 30 hours of professional development and then follow up with 15 hours of professional development at each of the secondary institutions, totaling 60 hours of professional learning. The Urban Advantage workshops were designed to both model museum learning experiences and allow teachers to learn or reinforce the underlying science content about the museum halls and as connected to the school curriculum, and the extended hours afforded ample time for teachers to engage in their long-term investigations to better implement exit projects with their students. Since the focus of the Urban Advantage initiative was engaging students in scientific inquiry, many workshops were designed around investigatory questions. Violet used many of the same questions in the workshops and adapted others according to the learning objectives she wanted to accomplish during the museum visit. Like the teacher workshop, Violet had her students rotate amongst the three Earth Science halls to make observations and gather information. In doing so, she adapted the museum's resources to create expanded affordances for her students' learning.

Learning Cultures and Affordances

Both schools and museums have distinct learning cultures—semiotic systems or schemas and corresponding practices (Sewell, 1999). There are different ways of learning in both spaces, depending on the available resources. When one learns, one also enacts culture as one becomes increasingly familiar with the schemas available. This is not to say that the cultures of museums and schools are without differences and nuances, but there are overarching characteristics that define learning interactions in each of these categories of institutions; for instance, the learning cultures of natural history museums are shaped around interactions with objects and exhibits, most of which are visual with minimal touching and manipulation. Urban Advantage formed the interface where the learning cultures of schools and informal science institutions meet.

Learning cultures are spaces where people access specific schemas and resources to meet learning goals, whether predetermined, designed, or self-articulated. The schema and resources constitute affordances, the relational, actionable, tangible, and tacit things available to a learner that allow for action (Gibson, 2014). Affordances that are enacted multiple times become embodied as practices.

The differences between formal and informal learning cultures are defined in terms of how the affordances are shaped and structured (Mocker & Spears, 1982). Formal learning describes what happens in schools, including K-12, postsecondary education, and professional certificates. With formal learning, the schemas and resources, including the learning objectives and approaches, are predetermined with minimal input from the learner. For example, we could think of learning standards and standardized assessments; when learning is designed with these outcomes in mind, there is little consideration of learners' interests and identities. In informal learning, the objectives are predetermined. In museums, exhibits and corresponding educational materials are designed by museum staff and/or consultants to convey particular messages, learning objectives or context and are usually somehow connected to the institution's mission. The learner's agency lies in how the learning is approached, as it is usually self-directed. There is also nonformal learning, where the learner has the greatest degree of agency in deciding both the learning objectives and approaches; this is more aligned with everyday conversations and activities in relation to a learner's community, culture and lived experiences. However, we can see from these descriptions that they are not absolute; learning cultures fall along a continuum from informal to formal. Furthermore, affordances are not fixed; rather, they constantly shift and transform as different learners access these affordances and somehow transform them to meet different goals, including developing, confirming, and maintaining identities (Figure 3.1).

The learning cultures of museums are shaped by the historical and contemporary societal cultures within which they are situated. From this perspective, we understand that while it is critical to understand the coloniality embedded in the history of museums, this coloniality is a part of the museum schema, which then shapes the learning cultures and influences the possibility of learning. However, with our human agency and as critically oriented educators, we can make this history visible and transform the learning culture to meet classroom and societal goals of equitable and just science education. While museums may seem static because of the physicality of the objects and displays, they become dynamic in how we use historical knowledge and critical approaches to desettle the existing displays. Beginning with questions such as: "Why is that object/artifact displayed in that way and with those other objects/artifacts?", we can start to engage learners in critical questions that can change how museums are viewed and how objects are interpreted and used in science teaching and learning.

Learning cultures are structured around schemas + resources ◊ affordances. Through learning, learners develop practices to access the affordances available to meet their learning goals. Table 1 offers examples of schemas, resources, and practices located in museums and other ISIs. It is important to note that although these are described as separate categories, they are not discrete categories but are relational. For example, schema influences the resources that are available and the practices associated with them. However, human agency can change how the schema is considered and how practices might unfold in relation to the changing perceptions of the schema.

Affordances and practices are relational. Even with seemingly static things like dioramas, how one learns, teaches, and interacts with the diorama changes as one becomes more familiar with the surrounding affordances and brings their own affordances into their relationship with the object. With

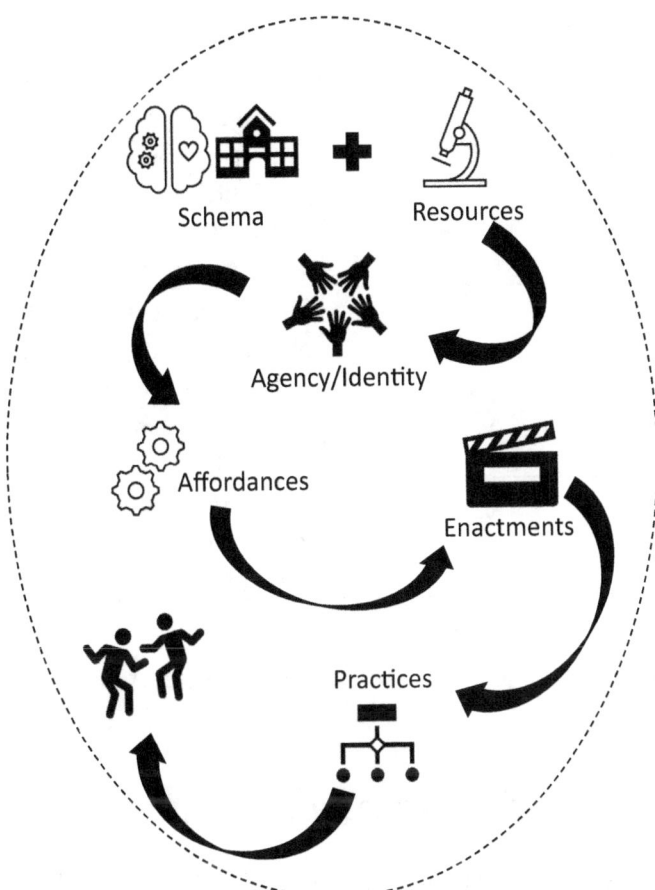

Figure 3.1. In the expansivising space new learning cultures are formed and identities confirmed/evolved as resources and schema from different learning cultures, in this case ISE and the formal classroom, come into relation. Human agency and identity shape the affordances and the ways that the affordances become enactments (immediate actions) and eventually repertoires of practices (patterns of enactments).

effective science teaching and learning being resource-dependent, in other words, having access to the tools, objects, and artifacts of scientific inquiry, museums have been positioned as enriching school science through access to these kinds of resources. Throughout this book, resources will be the most oft-used term to describe the affordances of museum/ISI-based learning, as these are the most visible and defining structures that distinguish these settings from formal classrooms. Furthermore, having access to science-related physical

Table 3.1. Examples of schema, resources, and practices.

Schema	Mission and ideology of the institution; beliefs, values, notions of how to conduct social life in a given setting; pedagogical approaches fostered in an institution.
Resources	Physical context (including objects and exhibits), intellectual resources (science content and research), symbolic entities (like history).
Practices	What emerges in encounters between schema, resources, and learners.

resources is crucial to engage students in authentic practices and experiences of science.

Affordances and Identity

Learning is integral to social life, and the complexity of social life compels us to view the relationship between museum learning and identity in a relational way. This avoids a cause/effect dichotomy that is not useful when describing the different learning cultures in which one participates and develops corresponding identities across their lived experiences. Examining agency/learning/identity relations allows us to describe how learning interactions with the museum and objects can mediate teacher identities. As learners become more familiar with the resources, schemas, and practices of a learning culture, they can also contribute to the transformation of the culture, which becomes a salient identity-shaping process. Agency and identity are dialectically related, so our power to act with given affordances mediates our identity development.

Learning to use, modify, and transform affordances expands a teacher's agency and not only transforms practices in the classroom but also the museum. This expanded teacher agency, in turn, expands the affordances available for students' science learning. With teacher learning that centers equity and the needs of diverse learners, the expanded affordances contribute to equity and broadened access to the learning culture of science. Equity and justice-oriented frameworks could be learned through usual teacher learning approaches, such as readings, reflections, and discussions, but also stemming from a teacher's own lived experiences with inequity and oppression.

Schemas and practices are also dialectically related (Sewell, 1992); as one learns the culture of a field—the schema and resources that shape the field—one can change practices to achieve goals within a given field. As one gains the ability to change and transform practices, one gains agency in a learning culture. We could also consider a radically conditioned agency where teachers can "reflexively and critically examine their conditions of possibility and in which they can both subvert and eclipse the powers that act on them and on which they act" (Davies, 2013, p. 426). This is the kind of agency necessary to transform science teaching and learning towards practices that rehumanize students and science learning. This is a *restorative agency*, the day-to-day thoughts and actions aimed at catalyzing the transformation of institutional and societal structures toward achieving an equitable and just society and planet. Like identity, the development of restorative agency is ongoing and starts with small actions. For teachers, these are the thoughts, actions, and decisions that are made in the act of teaching, design decisions in the design of activities and curricula and the longer-term actions of professional development and alignment with others and initiatives that center social justice and equity in science education. This sense of restorative agency can be deliberately designed in teacher education and emerge within teachers' collaboratives, where they are able to discuss salient issues in their classrooms and collectively think about and plan for ways forward.

In revisiting the vignette at the opening of this chapter, Violet was situated between learning cultures in the expansivising space between the museum and the science classroom. As she learned to use the museum resources, she extended her learning in her classroom by having her students experience the same learning approaches that she did during the teacher workshops. During the Urban Advantage workshops, she had access to the resources—museum educators and scientists, objects, and behind-the-scenes knowledge—of the natural history museum. She learned the schema, how to use the resources, and how to enact practices related to the schema and resources with her students. While doing this, she was expanding her agency as a user of ISI resources. She also became a resource for her students in ways that enabled them to access the learning culture of the museum, knowing how to look for and make connections with the scientific information presented in the exhibits and artifacts for their own learning. This allowed them access to science learning beyond their classroom, thus expanding their notions of science in terms of where and how it is practiced.

Culture Shapes Affordances

The culture of museums, in this case, a natural history museum, is structured around heterotopias, "combinations of different places as though they were one." (Kahn, 1995, p. 324); they are designed around finding the relationships between objects that may or may not be contextualized as they are as in their natural existence. For example, one display based on systematics—the science of describing and classifying living things based on genetic and morphological evidence, shows the evolutionary relationships of living things; it is called the Tree of Life. On this tree, model and preserved representations, starting with prokaryotes at the base, are arranged to show how they are related to one another and all life on Earth. However, these organisms do not *live* in these relationships, and although they may be right next to each other on the Tree of Life, they may never actually physically interact in their lifeworlds. Heterotopias also exist in zoos and botanical gardens where living specimens are on display. For instance, plants of the dessert might be displayed alongside each other—a succulent from South Africa alongside one from Arizona—but they are not found endemically together in lived geography.

Objects in museums are displayed in such ways as to represent certain scientific concepts—in the case of the zoo or garden, it may be animal and plant adaptations. They are also organized in ways representing different ideologies or worldviews, in most instances, those being the dominant views of Western Modern Colonial Science, as discussed in Chapter Two. To effectively use these spaces as science teaching and learning resources, it is important to learn the concepts behind the development of these heterotopias to contextualize the objects and not reproduce the embedded coloniality of the displays in teaching and learning. For example, the Tree of Life could be understood as one representation of life on Earth and even one interpretation of this object of systematics. However, we can also recognize that there are many other valid ways of categorizing and representing life on Earth and describing interrelationships of human, nonhuman, and material aspects that are also culturally embedded and resonant with diverse worldviews. These heterotopias or exhibits help to shape the learning cultures in natural history, botanical gardens, and zoos. When learning how to effectively integrate these exhibits in science teaching and learning, teachers learn not only the content but also the context—the systems and practices that were used to create the halls, including the historical context of some of the older halls and exhibits as well as best practices for using the objects on display in science teaching and learning. In essence, they learn about two nested cultures: the culture of science as practiced in a museum and the culture of using museum exhibits and resources to teach science. The Urban Advantage teachers thus developed culturally adaptive practices that allowed them to make connections between the museum's affordances and the teaching and learning goals of the schools and classrooms.

In the museum, teachers enacted a learning culture around teaching science through inquiry with objects. Paris and Mercer (2002) refer to this as a transactional model of interaction with objects, which is defined as "an object-based epistemology that transcends the actual object by virtue of the... social experiences engendered by the object" (Paris & Mercer 2002, p. 402). Agency and corresponding identity arise from these transactions, as learners use these objects to connect with "features of their personal lives, both actual and imagined selves, during their explorations of objects and museums" (Paris & Mercer, 2002, p. 402). With the ongoing development of a science teacher identity, science teachers explore and access objects in ways that allow them to further develop and confirm their teacher identity. If we extend this to an imagined self, accessing the learning culture of a museum also shapes the identities of the teachers they would like to see themselves become. Identity is fluid and is continuously reproduced and transformed during the activity of learning. Identity is also about recognition, "when any human being acts and interacts in a given context, others recognize that person as acting and interacting as a certain 'kind of person' or even as several different 'kinds' at once" (Gee, 2000–2001). Because of the fluidity of identity, "the kind of person" one is recognized as 'being,' at a given time and place can change from moment to moment in the interaction, can change from context to context, and, of course, can be ambiguous or unstable." (p. 99). Learning, identity, and agency evolve in relation to one another; thus, as museum-based resources and objects become part of the teacher's science teaching and learning activity, they also constitute the identity of a science teacher.

Use-Modify-Create and Agency

Learning, identity, and agency influence how teachers access and adapt the resources they encounter in museum learning to meet the needs of the classroom. The Use-Modify-Create (UMC) is useful for describing teachers' trajectories, from encounters with informal science resources to full integration and transformation of teaching practice. UMC describes a progression for youth engagement in computational thinking. It is based on the premise that structuring increasingly deep engagements allows students to, over time, gain increased agency in the skills in the development of computational projects (Lee, Martin & Apone, 2014) (Figure 3.2).

Through the process of being a "user" and then actively engaging with modifying and creating affordances, teachers develop increasing agency and begin to see how they can integrate both classroom and ISE schema and resources to transform STEM learning experiences and opportunities for their students.

Use—This means first learning the available affordances and corresponding practices of the museum. This includes learning about the history and content of the displays, objects, and dioramas, as well as other resources such as videos and digital displays. While they are learning about the affordances, they also learn the corresponding practices associated with learning in museums. Teachers begin to get ideas about how to enact resources in the classroom through modeling from museum educators and/or through brainstorming and dialogues with peers.

Modify—Faced with the realities of the classroom, teachers adapt the resources to fit the needs of their curricula and the diverse learners in their classrooms. It is an ongoing reiterative cycle of adapting, implementing, reflecting, and transforming, each time with practices becoming more embodied and reflexive. This is also where some teachers may experience resistance from administrative mandates and schooling structures, such as standard assessments that create challenges to enacting ISE-learned teaching practices. However, through the modification process, teachers find ways to adapt resources to create affordances they feel are necessary for meaningful science learning. This modification process is catalyzed by teachers engaging in dialogues with other teachers and educators, where they discuss challenges and exchange ideas about how to adapt resources while considering the challenges. This is also an opportunity for developing an expanded restorative agency when the learning needs of diverse students are central to the discussions.

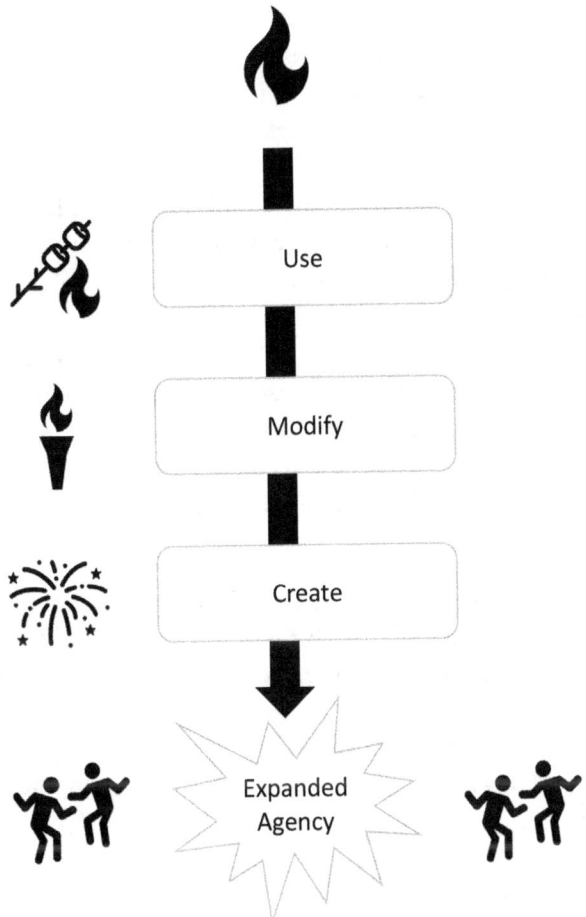

Figure 3.2. Applying the UMC model (Lee et al., 2014) to teacher learning. Each enactment through the UMC cycle contributes to establishing repertoires of practice, expanding teaching agency and corresponding identity.

Create—Teachers become bricoleurs in creating new affordances from the ones they initially encountered in the institution, along with other affordances and practices that they learned in other spaces and in practice. Through the UMC process, teachers also develop expanded agency and corresponding identities in relation to the affordances and new ways of teaching and learning. Learning experiences are structured so that learners first learn how to *use* existing affordances before *modifying* them to achieve a design outcome or educational goal. After learning and successfully modifying affordances, learners move towards being able to *create* new affordances that, in turn, transform learning culture. As learners go through this process, they develop increasing agency in their knowledge of the applications of the affordances and use them to meet different goals. The process is similar to teachers learning to use museums, science-rich cultural institutions, and other resources. Although learning processes are nonlinear, I will describe them in a linear fashion here in order to elucidate the relationship between learning and agency. We could also think of it as a person with the identity of a science teacher encountering a museum for the first time.

As teachers learn the affordances and develop practices, teachers begin to make modifications to transfer their teacher-learning experiences into their classrooms. Often, teachers have positive learning experiences in informal science institutions and are excited to recreate similar activities with students, but in doing so, they will undoubtedly have to modify what they've learned and experienced. In doing so, they consider:

- Who are their students as learners, including what are their social identities, and how does it position them vis-à-vis science?
- What is the feasibility of field trips? This depends on questions of location and transportation, school policies regarding field trips, costs of field trips, etc.
- What are some of the places closer to my school that would allow students to have a similar experience?
- What are some of the practices that I learned that would help my students to learn in or near the classroom?
- How can I adapt what I've learned to meet the needs of students in my classroom?
- How can I create new resources and science learning opportunities for my diverse students?

In workshops that structure in time for reflection and conversation amongst participants, the modification usually begins while teachers are in the museum and prompted to reflect on their students and discuss how they would use or modify the resources that they encountered and experienced during the professional development. The process continues when they return to their schools and classrooms and adapt the affordances and practices in the context of available affordances in the classroom. Teachers then begin the reiterative process of transforming learning culture as they both *modify* and *create* new learning cultures for their students.

Objects Create and Shape Museum Affordances

It is the mission of science-rich museums to communicate the work of scientists to the public using objects and visual displays. Objects in museums are imbued with value because of their place in a museum and because of what they represent to people who interact with them. This also means that these objects are not neutral but have histories behind how they came to the museum and the knowledge they represent to different people who encounter them. To a scientist, an object could represent objective scientific facts or proof of scientific phenomena. To a teacher, although the facts may be present (in the accompanying text), the same object could represent an object of inquiry that is subject to interpretation. To a student coming into the museum, objects are polysemic—they are based on the knowledge and schema that students carry with them and their meaning-making as they interact with the object grows from this knowledge base. For example, a meteorite could be a giant rock to a student whose contact with such specimen are boulders that they climb on in local parks; however, with a teacher as a mediator, the student can begin to see that this "giant rock" as an object that has provided information about the composition of our solar system. To students, the rock is now schematically a meteorite and has a specific role in scientific understanding and knowledge production. The students could also learn that this meteorite is a sacred entity to a particular group of Indigenous people, including stories that they tell about the origins and significance of this meteorite as well as the colonial history of how the museum acquired this object. This begins the discussion of the coloniality embedded in museums and scientific knowledge production, including harms done in collecting artifacts and efforts at repatriation. This allows students and teachers to view these objects with multiple lenses while expanding their schema to include a more critical analysis.

What Do You Wonder?

Teachers use objects in a museum to enact inquiry-based teaching change and adapt the meanings of objects to meet pedagogical goals, affordances, and constraints in the classroom. One of the things that connect classroom and museum learning is the use of questions to shape investigations. In prior literature, this was termed an "investigatory stance" (Hapsgood & Palinscar, 2002) to object/inquiry-based learning, whereby a learner poses questions in relation to the object to spark deeper investigation of both the object and the context of the object. Teachers learning to teach in a museum are taught how to make these interactions with objects relevant to meeting their pedagogical goals, including science content process learning and social goals of equity in science. In doing so, a teacher learns how to appropriate the museum's resources to enact a pedagogy around objects and inquiry. When teachers have access to various fields or learning cultures, they also have access to more resources and opportunities to learn about different cultures. The knowledge of objects—what they represent—and the ability to teach with objects reproduces the structure of the museum as a site of science education and research. Using objects to enact inquiry-based science becomes patterned actions of science teaching and learning in the classroom, becoming a resource for all museum teaching and learning participants. The knowledge of using objects to teach can also be transformative in the classroom—the teacher embodies the identity she gained in the museum and returns to the classroom where she re/creates the structure in the classroom to enact a similar object/inquiry-based pedagogy. Creating new structures in a field is agentic and potentially will change the identities of teachers and students, as the re/produced structure allows a teacher to re/produce an inquiry-based teacher identity in the museum and in the classroom. In the following sections, I use empirical vignettes to describe the progression of the learning culture that developed in the expansivising space between the ISIs and schools.

Structuring Encounters with New Resources

Engaging in a new learning culture is an opportunity for the expanded agency. However, when encountering a new learning culture, there can be conflict as one is adapting to this new culture in preexisting contexts. In Urban Advantage, middle school teachers—many from schools under-resourced in terms of science resources—were afforded access to new resources inside and outside the classroom. They received equipment to extend museum investigations into the classroom as well as new pedagogies for engaging students in science learning in cultural institutions. While the teachers expressed excitement and enthusiasm about having this new access, many were faced with the challenge of utilizing the resources, as there were many challenges to implementation.

During the first year of Urban Advantage, many of the first-year teachers struggled with issues ranging from the logistics of planning field trips to the process of integrating investigation-based pedagogy into their classroom practice. It was a struggle of conflicting learning cultures in many instances—the newer classroom teachers did not yet have the schema, both classroom and the museum, necessary for the successful integration of resources. In addition, a number of the teachers were not yet familiar with project-based assessment or hands-on inquiry-based learning.

Negotiating Different Learning Cultures

The teachers needed enactment structures in order to put into practice their professional learning with their students. Urban Advantage started with the notion that the museums would teach teachers content and how to utilize the museum resources, while the Department of Education would provide training and support for the project-based implementation—the classroom and curricular aspects of the program. In response to their role, the DoE hired teacher facilitators—teachers who were experts at facilitating project-based assessments with students—to consult with the partner institutions and mentor teachers in enacting project-based learning in their classrooms. Although they were able to assist teachers in the classroom implementation of the Exit Projects, they were not familiar with the learning cultures of ISI cultural institutions and were therefore not able to offer teachers the help they needed in accessing the resources of the institutions to do the Exit Projects with their students. One teacher noted, "He knew his stuff about Exit Projects, but every time I asked him if my kids could do this question, he would say no, no. They wanted to design an exhibit, but he said no. We talked about exhibit design in the workshop, so I allowed my students to do that project anyway." Other teachers told similar stories. The facilitators were experienced and effective classroom teachers; however, because they did not participate in the Urban Advantage professional development sessions nor did they actively practice using ISIs to teach and learn science, they were ineffective in bridging the learning cultures between the ISIs and the classroom.

Although science teachers expanded their agency by participating in new learning cultures through professional development, they were faced with challenges in the classroom. For example, while field trip activities were modeled during workshops, some teachers were challenged to design their trips that were both relevant to implementing the Exit Projects and meeting their curricular goals. Teachers were unfamiliar with developing and implementing activities that could bridge the museum and classroom environments. For example, early in the initiative, I observed a Urban Advantage teacher and his class in the museum halls, and I did not see evidence of students engaged in any guided or structured inquiry with the exhibits. I suggested to the teacher (and the students) that they observe the objects and displays and come up with questions about the objects with the prompt, "What do you want to know?" This helped to focus the students for a moment. However, I wondered in what ways the experience would be built in the classroom. Although we modeled generating questions from observing objects and displays during the workshops, this interaction with the teacher made me realize that not every teacher was able to make a full transference; however, the field trip with students was a start. Various factors, including time, administrative constraints, and lack of opportunities to engage in dialogues with other teachers about different ways to enact what they learned with their students.

The Urban Advantage teachers needed examples and activities to immediately implement with their students. They also needed peer support for the long-term planning and enactment of using museums to complete the Exit Projects and realize science teaching and learning. Effective teacher facilitators would have to be familiar with the museum's resources and the pedagogy of using museums to teach science. There was a need for an intermediary field to bridge the cultural gap between the museum and the classroom. This sentiment resounded during an end-of-year focus group session where a teacher noted, "What we need are institution teacher liaisons… [to say] here is what we really need as teachers…content is good, but applicability is paramount—the institution has to be presented as more than or beyond a resource."

To bridge the gap between the learning cultures of the museum and the classroom, a group of teachers who experienced the full 60 hours of professional development in the first year of Urban Advantage self-selected (via application) to become peer support for teachers who were new to the initiative. Ultimately, seven teachers were selected to become Lead Teachers; their selections were based on their attendance and active participation in the teacher learning activities and end-of-term science fairs, as well as the relationships they developed with the cultural institution partners. While they had varying experience with using out-of-classroom resources for science teaching and learning, they were all new to using the museum and partner ISIs for specific curricular goals with their students.

Modifying Resources/Creating Affordances

Urban Advantage first-year teachers, including the Lead Teachers, faced both challenges and successes in making effective use of the resources to do the Exit Projects with their students. Pamela mentioned as her reason for wanting to become a Lead Teacher, "I fully understand that many science teachers would love the opportunity to have other teachers…someone to help guide them. I know I would have wanted that when I was new. The students shouldn't suffer because of our lack of knowledge."

Pamela was in her early 30s and had been a science teacher in NYC for seven years. During her first year in Urban Advantage, she not only had to negotiate to be able to integrate new resources into her practice, but she also had to manage to implement independent science investigations while having to cover her science curriculum. She had to become familiar with the museum's structure, learn how to enact teaching using the objects and exhibits and extend that object/inquiry-based pedagogy into the classroom. After going through the first year, Pamela knew firsthand the challenges that a classroom teacher would face in being successful with the initiative—including time management, managing student activities during field trips, and negotiating standardized tests. As a Lead Teacher, she would have the opportunity to share her learning experience with other teachers and help them to learn and enact the inquiry-based culture with their students.

In stating her desire to "help guide" other teachers, Pamela has presented herself as a teacher who is familiar with the museum-based pedagogy and the structure of Urban Advantage and would be able to share her practice with others. Similarly, Violet, another Lead Teacher, stated, "I decided to become a lead teacher to encourage other teachers to use the resources [of the institutions] to guide them through the stumbling blocks I encountered and to help them understand how best to use 'out-of-classroom' scenarios." She also used the word "guide" in her discussion. Pamela and Violet, as mentioned at the opening of this chapter, both recognized the need for enactment structures for new Urban Advantage teachers. Both teachers wanted to be a part of that enactment structure—to share what they have learned during their first year in Urban Advantage in a peer leadership capacity. They both developed a sense of agency in negotiating the structure of the new pedagogy and resources during the first year. They implemented the field trips and had several successful student Exit Projects that demonstrated science inquiry using the cultural institutions. Pamela and Violet felt that they were able to help others, perhaps believing that their experience could "jump-start" other teachers' experience—that they would not have to, in essence, start from a place of total unfamiliarity.

During the second year, the Lead Teachers were still users of ISI resources; however, they moved to modifying ISI resources by working alongside museum educators and scientists in designing teacher learning experiences. The teachers, museum educators and scientists began to learn and create shared practices that merged ISI and classroom affordances. Each cultural institution had its expertise, and these areas were equally valued (for the most part) towards the goal of meaningful ISI-based teacher learning. Within this context, Lead Teachers learned together how to support other teachers in their journeys towards agency in ISI-related teaching and learning. In essence, they helped to create a learning culture that was in between the cultural institutions and the classroom.

The overarching teacher-learning goal of Urban Advantage was for teachers to be able to effectively use, modify and then even create ISI-related affordances for student learning. The ISIs were to be contextualized as extensions of urban middle school science classrooms. As such, the Lead Teachers were positioned to change teacher learning but also to transform science education city-wide. Each of the Lead Teachers was partnered with one of the Urban Advantage cultural intuitions based on the location and/or content expertise of the teachers. From there, they worked with the cultural partners to align the key activities of Urban Advantage with classroom curricula and assessment goals.

Negotiating Identities in Expansivising Spaces

At the first Lead Teacher meeting, flipchart papers were placed around the room with questions for the Lead Teachers to brainstorm—this was done both as an icebreaker and to field the collective ideas in the room. One of the charts was headlined, "What do you see as your role as Lead Teacher during the upcoming year?" This chart became key in articulating the Lead Teacher roles and was turned into a document called a "memoranda of agreement."

As a manager in the Urban Advantage initiative, it was my role to facilitate the Lead Teacher group. When I started the Lead Teacher group, I had preconceived notions of how the teachers would enact their leadership roles in the cultural institutions. I envisioned them easily slipping into co-facilitation scenarios with the cultural partners and acting as peer mentors for teachers new to the initiative. I thought the transfer from being a classroom teacher to being a peer leader would have been seamless, and the Lead Teacher sessions at the museum would focus on content and objects. However, after the first Lead Teacher meeting, through the evaluator, I learned that many of them were still unsure of how they would enact their Lead Teacher roles. As this was early in the initiative, it allowed a quick pivot from focusing on using museum/ISI resources to peer-sharing sessions about enacting their roles as Lead Teachers. The use of resources emerged in the discussions rather than the discussions being shaped around the resources.

I then planned and facilitated the monthly Lead Teacher as a forum to network—to share and discuss their experiences in working with the institutions and in their classrooms, openly struggle with issues around Exit Project implementation, and offer each other suggestions and advice. Having adapted a practice associated with community building from my work in experiential learning with Outward Bound, I facilitated a dialogic space that allowed the Lead Teachers to share experiences and ideas while building shared meanings and practices that identified them as a community. As a facilitator of the group, I tried to sit back and allowed discussions to unfold, with little input unless they needed guidance about policy, political issues, and the vision of the Urban Advantage initiative.

I was able to facilitate the Lead Teachers as they evolved through the UMC process and formed a collective while doing so. They were aware that I was also doing research, but a potential conflict of interest never came up as an issue. When I asked for permission to videotape, the participants were amenable as long as they were assured that I would not share the videotape or personal information beyond the group, as written in the informed consent that they signed. I participated in group discussions sometimes as the voice of the cultural institutions, other times as the voice of Urban Advantage central leadership and other times as the voice of a previous classroom teacher. Although I had previously taught high school, I felt that some of the issues that they struggled with were experiences that I shared as a teacher—such as having their own classroom (one that is not shared with other teachers) to display their resources without worry of them getting vandalized or not cared for.

Lead Teachers began to create a learning culture that was between the museum and their classrooms. Their learning culture was structured around their group interactions and discussions about enacting object/inquiry-based pedagogy and doing Exit Projects with their students. They also discussed their participation and interactions in working with partner institutions in planning and implementing the workshops for new Urban Advantage teachers. Two vignettes describe some of the challenges the Lead Teachers faced as they negotiated different learning cultures and policy structures.

Tensions in Learning Cultures

Because my aim was to establish a community of practice at the nexus of two learning cultures, I used Lave and Wegner's (1991) idea of legitimate peripheral participation to consider the role of the Lead Teachers in relation to the museum and other ISIs. This approach describes how novices learn and move to the center of the practices as a learning community forms. The theory positions people either at the periphery/margins or centrally in a community of practice, depending on their skill set and contributions to the community. It is important also to consider the ways that people are positioned and/or position themselves at different points along the continuum from the periphery or margins to the center. At the center, there can be the desire to maintain power and keep it enclosed from the participation of diverse learners. Whereas at the margins, as bell hooks (1989) reminds us, there is more freedom, which means expanded opportunities to create a range of new affordances from resources at hand. It is a space where learners can express the most creativity as they brainstorm, experiment, and generate ideas for action. However, resistance often occurs as these new ideas move towards the center and before transformation takes place. In the case of the Lead Teachers vis-à-vis the cultural institutions, this means that the professional development could not continue as it has been, and these tensions became evident once the Lead Teachers began asserting their roles and identities not only in the cultural institutions but at times in their own schools. For Urban Advantage, where the goal was integrating the cultural institutions as resources for middle school science teaching city-wide, this meant shifts in the schema at the center in order to make space for new affordances to emerge. As I will later describe, this event became a pivot for shifting the center.

Tensions of Identity

Jake's partner institution had what he described as "pre-programmed professional development"—it was fun and interesting, but the activities were not relevant to the Exit Projects. Jake expected the professional development to be tailored to the needs of the Urban Advantage teachers; the activities offered should have been directly related to implementing the Exit Projects, but they were not. He struggled with setting up planning time with the educator from his partner institution; he felt that this educator was evasive and unwilling to include others in the planning process. "I tried to call, email and meet with them but could not get through, I think she [instructor] is on vacation or something." I attempted to act as a go-between—emailing the museum instructor and trying to set up planning time, but my efforts also went ignored. He was visibly upset when he discussed his futile efforts to make himself available to the museum educator for planning.

Jake's agency was truncated as he attempted to fully participate in the professional development planning. He had a lot of ideas about changing some of the activities to match what teachers needed to know and do for the Exit Projects. Having participated in last year's professional development, he believed that he had valid feedback, and he was excited about the possibility of affecting change for this year's teachers. He felt that his position as a Lead Teacher would afford him a certain degree of input in the planning of professional development; however, it seemed that he was kept at the margins vis-à-vis his cultural institution.

I was not present for any of the interactions between the museum educator and Jake, so I could only base my interpretations on what I learned from Jake and on my own interactions with the museum educator. Although she never made any direct reference to specific events in her conversations with me about her interactions with Jake, she often described him as a "character." I know that Jake has a big personality because of his magician-performer identity, and he is confident and passionate about his teaching skills. Jake also admitted that he can be "undiplomatic" in stating his opinions and was very critical of the museum's professional development agenda. It became obvious to me that there were tensions in role negotiations and ownership of the professional development agenda. Some of the cultural institutions viewed the Lead Teachers' roles as more supportive rather than collaborative, whereas the Lead Teachers aimed to work collaboratively; however, recognizing the value in their role of aligning the professional learning activities with the realities and requirements of the classroom. If the project-based assessments were going to be successfully implemented, there must be a meeting at the center of all participants involved.

Partnerships between individuals from various learning cultures can be fragile as people negotiate their identities and roles. My role as the facilitator of the Lead Teacher group was to also serve as a liaison between them and the museum educators. When issues such as this arose, I did my best to schedule time for all participants, including myself, to engage in dialogues to generate shared understandings and goals moving forward. A dialogic structure allowed for a discussion of issues at hand and agreement upon actions to reduce arisen tensions (Bayne, 2009). While I was able to do this within the Lead Teacher group, it became more challenging to include different participants due to scheduling and location. By the time I was able to talk to the museum educator about this incident, I had asked an undirected question: "How are things working out with the Lead Teachers?" The museum educator replied that everything was working well, which led me to believe that the issue between them and Jake was resolved, and Jake did not seem to vent about it in subsequent Lead Teacher meetings.

In contrast to Jake's experience, Mike mentioned that he had made significant contributions to the planning and implementing the Urban Advantage sessions. Mike cited that his institution was "easy to work with and receptive to my suggestions in planning the [professional development] for teachers." Mike was able to enact his classroom teaching culture by aligning the museum's resources and professional development activities with what Urban Advantage teachers needed to know to implement the Exit Projects. In Mike's case, the center seemed more distributed, allowing the margins to appear blended and not form a strict dichotomy or even gradient. Mike and the partner museum educator brought their respective spheres of expertise in relation to each other. Mike's practical applications to the classroom were valued and a necessary contribution to the enactment structures that the Lead Teachers were creating for other teachers. The museum educators who worked with Mike described how helpful he was to them in reviewing their professional development activities and offering suggestions for better alignment with the middle school curriculum and Exit Project requirements.

The tensions experienced by some of the Lead Teachers are what happens when ideologies and axiologies clash in the expansiving space. However, tensions are necessary in expansivising spaces to negotiate and shape the new affordances being generated so that they are reflective of all participants. In these tensions, new individual and collective identities are also formed.

Science object/inquiry-based education was shared practice between the cultural institution and the Lead Teachers—it adapted and expanded from being a museum-based pedagogy to being one that was used in Urban Advantage classrooms. While there was a common understanding of object/inquiry-based learning, it created tensions when deciding how to approach teacher learning.

As a Lead Teacher, Jake was positioned as the "expert" in classroom teaching and knowing what was needed to be successful in the implementation of teacher education in the classroom, he wanted to move to the center of the planning of the professional development sessions. However, in being not fully central in the cultural institution and not having access to the resources—in this case, the instructor/facilitator—he had to find other ways to enact his expertise in the cultural institution field. This demonstrates that even within expansivising spaces, it is always necessary to be mindful of power differentials and manage imbalances. Jake did not have as much power in the cultural institution as he did in his classroom. The role tension he described occurs when people are exposed to unfamiliar schemas and practices and compelled to change—there is either resistance or transformation.

Jake's experience became a shared experience in the group as it enabled discussion amongst the Lead Teachers about how each of them was working with their cultural partners and allowed the group to process successes and challenges in the context of role definition. These stories and shared repertoires became important in shaping the identity of the group and the individual Lead Teachers—these shared experiences defined their identities in relation to others both inside and outside of the Lead Teacher group.

Tensions of Power

Although schools often experience space challenges, during my visits to different schools, I have often found places like science laboratories that remain unused. Pamela's school had one such room that was not touched, according to school lore, for 30 years. With her Urban Advantage resources and role as Lead Teacher, Pamela saw this as being the space to establish a demonstration classroom. Pamela and her colleague Andrea negotiated with the principal to have this room for a science resource room. They also wanted a space where science teachers could plan and prepare their lessons and science clubs could meet after school. This was the old science prep room, so it seemed appropriate that it should become the "new" science resource room. "We cleaned out the whole room and got rid of a lot of old chemicals and stuff." I pictured how the room might have looked before—grey, dust encrusted with old beakers and flasks filled with crystallized chemicals. I also thought about Pamela and Andrea having to contend with the noxious chemicals during the oppressive summer heat, but being familiar with the system, I knew that if they wanted it done and had to wait for the red tape to be cut, it would not have gotten done and the room would have remained useless for another 30 years. "We had them [janitors] paint the room, and the kids made signs," even though it was after dark when I saw the room, it was bright yellow with a few potted plants, which gave the impression of the high noon sun. The counters were clean, and the covered microscopes were lined up in neat rows towards the window end. There were cheery signs ornately handwritten by middle school students that labeled the cabinets, along with photographs of the students who frequented the room.

I asked Pamela what gave them the impetus to clean up this room. She responded, "Well, I thought if we were going to be Lead Teachers, we should have a space for the science resources." Her role in the learning culture she was co-creating with the other Lead Teachers warranted her having a space that demonstrated affiliation with that culture. Her school had a mock courtroom for humanities, so Pamela felt that it was only fair that her school also have a space dedicated to science. Leveraging her Lead Teacher role within the school, she acquired the room and involved other staff and students in transforming the room, thus affording a sense of collective ownership of the room as a resource for science education. Science teachers in her school now had a space to meet, plan and discuss science. The room also created new affordances for students to expand their interests in science beyond what they learned in the curricula and formal school space.

Pamela's was the first school the Lead Teachers visited as a group. Since one of the roles that the Lead Teachers generated for themselves was to create the demonstration classrooms, the group decided to have the meetings at a different Lead Teacher's school each month to see how the others set up the Urban Advantage resources in their rooms. Intervisitations enabled the formation of a collective identity around classroom display, as described in Chapter Four. The display of objects in ways that motivated inquiry became a central practice of the Lead Teacher group.

Enacting new cultures in existing spaces does not come without tensions as the borders within become awakened. Pamela was the lead science teacher in her school and managed to get a science resource room, create her own science inquiry classroom, and support the science and teaching-learning efforts of the other science teachers in her school. Pamela did this as her school was changing principals, an event that sometimes creates the space for initiatives and practices at the margins to claim spaces towards the center.

At first, Pamela's new principal was not involved with any science education initiative, including Urban Advantage. According to Pamela, because math and English language arts were mandated assessments, her energies were focused on those areas. As such, Pamela served as the ad hoc administrator for the Urban Advantage and other science teachers in her school. For example, Urban Advantage had an annual event to update administrators on the project and usually featured invited officials from the Department of Education and City government, a good opportunity for visibility for principals and other school leadership. Pamela attended for her principal. Pamela also registered teachers and communicated professional development information regarding Urban Advantage within her school instead of her principal. "I wanted to make sure that my school stayed involved," Pamela said when she discussed her reasons for "taking over" in response to her disinterested principal. As Pamela's teaching identity was becoming more aligned with Urban Advantage along with her leadership role therein, she did not want to risk her school's involvement in the initiative. However, as the principal's role within her new school settled, the struggle for power over the Urban Advantage ensued.

The principal mentioned to a program evaluator that she "felt out of the loop" in regard to Urban Advantage; she started to demand that notices about Urban Advantage be sent directly to her instead of through Pamela. From what Pamela relayed to me, she made every effort to keep the principal in the loop (by following up with program-wide emails sent to all Urban Advantage administrators), but when there was minimal response, Pamela used her agency within the school and within the initiative (as a Lead Teacher) to ensure her school's continued participation in Urban Advantage. As participation in the initiative was in high demand, a non-response from a principal would jeopardize a school's participation, a risk that Pamela was unwilling to take. From what Pamela relayed, her principal began to micromanage Pamela's activity with Urban Advantage both within the school and in the larger initiative.

Pamela felt that she needed support from Urban Advantage, so she asked me to reach out to her principal about her school's continued participation in a demonstration classroom. It seemed that her principal wanted to keep her center exclusive, "she will only speak to other principals and administrators," Pamela mentioned. I emailed her principal twice but got no response, and as it was approaching the end of the school year when Pamela made this request, I was unable to pursue this matter further. Additionally, as I had no power within the structure of the Department of Education, I would not have been able to influence any decisions that the principal could make about Pamela's role in her school. I would have only brought to the table the symbolic capital of being a part of the central leadership of Urban Advantage. Pamela was able to continue in Urban Advantage both in her Lead Teacher role and her demonstration classroom, but from what I understood, the principal took control over the latter, extending it as a professional development space for the assessed subjects. Pamela's experience demonstrates the tensions that happen in expansivising spaces where power exists outside of that space.

Coalescing Learning Culture and Community

The Lead Teachers had varying experiences within their schools and in partnership with the different cultural institutions. They also had different experiences that allowed them to be at the center and other times at the margins of different learning cultures and contexts throughout different aspects of the project; however, whether center or margin, they were able to evolve their identities, collectively/individually, to reflect their museum learning and resources that they encountered and modified to meet identity and teaching/learning needs. The Urban Advantage initiative expanded the science teaching and learning affordance that science teachers had access to; however, to effectively enact science teaching and learning in the museum and integrate those practices in the classroom, enactment structures were needed. This necessitated the Lead Teachers to reflect on how they were modifying the museum resources to create new affordances for their students. As they modified and created affordances for learning, they also created opportunities for expanded repertoires of professional development practices for cultural institutions. They also created tensions as they negotiated with structures that were outside of the expansivising space but wielded power over what was possible there for different individuals. However, the collective knowledge generation still afforded degrees of transformational changes in their schools and classrooms. The Lead Teachers became resources for other Urban Advantage teachers by sharing their knowledge and practices.

The experience of interacting with objects and learning how to teach inquiry-based science using objects enabled the reproduction of an inquiry-based culture of science teaching and learning within Urban Advantage, which became evident in the Lead Teacher group, as they created expansivising artifacts—such as the assessment rubric—that codified the inquiry-based learning that they learned in the museums.

The Lead Teachers developed a community of practice around creating a structure that allowed both themselves and other teachers to enact object/inquiry-based science in the museum and the classroom. Participation in the Lead Teacher group enabled them to continue to expand their agency by learning in the museum field and sharing/building their classroom practice of using the resources provided to them in Urban Advantage. As they were learning and participating, they continuously re/produced their identity around being object/inquiry-oriented teachers and becoming peer leaders in support of other teachers who were able to enact the inquiry-based pedagogy and Exit Project implementation.

Expansivising to a New ISI Partner

The New York Aquarium came on board as a new partner during the initiative's third year. They had to quickly get up to speed to be ready to begin the professional development sessions that would start in a few months. Elena and Jake met the new partners during one of the summer meetings and then spent a Sunday at the Aquarium reviewing their resources and co-planning the professional development sessions with the Exit Projects in mind. The Lead Teachers helped to facilitate the Aquarium's entry into the Urban Advantage learning culture. Jake and Elena suggested specific exhibits in the Aquarium and corresponding activities that met the goals of covering life science content and modeling Exit Project investigations. Jake was very proud of the transparent grid he developed, which could be placed on the glass of an exhibit to study animal behavior.

By co-facilitating teacher education sessions and consulting with partners, the Lead Teachers developed a learning culture that was situated between the cultural institutions and the classroom. This emergent learning culture served as an expansivising space for integrating and developing affordances for object/inquiry-based science teaching and learning. The Lead Teachers were able to extend the learning culture they developed in this expansivising space to other Urban Advantage teachers by providing examples and activities in classroom enactments of the initiative. As both individuals and as a part of a collective, the Lead Teachers built a group identity as peer leaders and inquiry-based teachers. This enabled them as individuals to transform the science teaching and learning structure of their schools through the re/creation of spaces and availability of resources for other teachers in their schools to enact inquiry-based teaching and use the museums as resources for teaching.

Notes

1   There were seven partnering institutions at the time of the research, now there are 8.

References

Bayne, G. U. (2009). Cogenerative dialogues: The creation of interstitial culture in the New York metropolis. In *The world of science education* (pp. 513–527). Brill.

Davies, C. B. (2013). *Caribbean spaces: Escapes from twilight zone*. University of Illinois Press.

hooks, b. (1989). Choosing the margin as a space of radical openness. *Framework: The Journal of Cinema and Media, 36*, 15–23.

Gee, J. P. (2000–2001). Identity as an analytic lens for research in education. *Review of Research in Education, 25*, 99–125.

Gibson, J. J. (2014). *The ecological approach to visual perception: Classic edition*. Psychology Press.

Hapgood, S., & Palinscar, A. (2002). Fostering and investigatory stance: Using text to mediate inquiry with museum objects. In S. Paris (Ed.), *Perspectives on object-centered learning in museums* (pp. 171–190). Mahwah, NJ: Erlbaum Associates.

Kahn, M. (1995). Heterotopic dissonance in the museum representation of Pacific Island Cultures. *American Anthropologist*, New Series, 97, 324–338.

Lave, J., & Wenger, E. (1991). *Situated learning: Legitimate peripheral participation*. Cambridge University Press.

Lee, I., Martin, F., & Apone, K. (2014). Integrating computational thinking across the K–8 curriculum. *ACM Inroads, 5*(4), 64–71.

Mocker, D. W., & Spear, G. E. (1982). Lifelong learning: Formal, nonformal, informal, and self-directed. Information Series No. 241. http://hdl.voced.edu.au/10707/116040

Paris, S., & Mercer, M. (2002). Finding self in objects: Identity exploration in museums. In G. Leinhardt, K. Crowley, & K. Knutson (Eds.), *Learning conversations in museums* (pp. 401–423). Mahwah, NJ: Erlbaum Associates.

Sewell, W. H. (1999). The concept(s) of culture. In V. E. Bonell & L. Hunt (Eds.), *Beyond the cultural turn* (pp. 35–61). Berkeley: University of California Press.

Sewell, W. H. (1992). A theory of structure: Duality, agency and transformation. *American Journal of Sociology, 98*, 1–29.

· 4 ·

MODIFYING RESOURCES: ADAPTING MUSEUM DISPLAY AND INVESTIGATION FOR THE CLASSROOM

Effective science teaching and learning necessitates access to a range of physical resources that serve as models for scientific phenomena and afford hands-on learning and deeper engagement in science both inside and outside the classroom. Realizing that field trips to the institutions alone could not facilitate quality investigations, Urban Advantage provided schools with tools and material resources to extend the investigations into the classroom. Each of the participating cultural institutions selected resources that would support the content of their professional development in the classroom. For example, the botanical gardens chose a grow tent so that classrooms could do controlled experiments with plants. The American Museum of Natural History selected rock samples to facilitate the teaching of the rock cycle and characteristics of rocks in support of Earth science-related projects. As science is a resource-intensive subject, the Urban Advantage resources aimed to help schools build their science teaching collections and to help teachers create spaces where science inquiry is an ongoing and visible practice. However, it is not simply a matter of providing resources but also affording teachers the opportunity to learn different ways to use, modify and even create their resources to support science teaching, learning and inquiry.

One of the roles of the Urban Advantage Lead Teachers was to use these resources to set up "demonstration classrooms." These classrooms would have the resources on display and model an ongoing practice of object/inquiry-based science learning. Knowing that different schools had different access to resources and administrative support, the Lead Teachers decided to rotate the monthly meetings at their schools to visit each other's classrooms to see how they set up and used the resources. Since my role was to facilitate these meetings, this rotation of sites also enabled me to see the different classrooms and experience the interactions of the Lead Teachers with each other within each other's created spaces.

Display as a Pedagogy/Display as Inquiry

In learning cultures, meaning-making occurs during various transactions and interactions between people, things and contexts. In informal science institutions such as natural history museums and botanical gardens, object observations and interactions are central to the learning culture. To support the expansivising of aspects of ISI learning culture in the classroom, each Urban Advantage teacher (and student) was given a special science notebook to record observations. This notebook had designated lines for writing and blank spaces for drawing so that both words and drawings of observations could happen on the same page. Teachers were prompted to model using this tool to look at objects and displays and write their observations and impressions in their notebooks. They were then encouraged to turn these observations into inquiry questions, i.e., "What do I want to know," that could lead to a science investigation. This placed the dialogue between teachers and objects as central to the learning or meaning-making experience in a museum.

Developing Museum Literacy

Objects are central to the learning culture of museums. Museum literacy is the ability to read objects (Bain & Ellenbogen, 2002), and reading or interpretation of the object is influenced by the prior knowledge, experiences, and identities that the viewer brings to the encounter with the object (Rowe, 2002). The interaction between the object and the viewer has two sides: "the interpretive framework brought to bear by the individual [as an active agent], which is both personal and social, and the physical character of the artifact" (Hooper-Greenhill, 2002 p. 112) and this creates a dialogue between the viewer and object. Objects can also be considered as something that creates a bridge between prior experiences and future interactions.

The schema and resources of a museum are structured around the display of objects through curatorship. Curators have the power to facilitate meaning-making and for visitors to make connections between the objects on view and their lived experiences (Bunch, 2021). Objects are situated in relation to other objects to communicate selected information about the object itself, the science-related content and processes in relation to the object and the object's significance to scientific endeavor. Learning cultures in museums are shaped by visual culture that centers on social practices of looking and seeing as a way of learning and knowing (Hooper-Greenhill, 2002). Teachers learning to teach in a museum participate in this visual culture, which is nested within the learning culture of museums. This is a form of museum literacy that teachers could transform and enact in their classrooms to meet related goals. This could be both a source of agency and identity for teachers since it affords them access to different ways of knowing and enacting science teaching and learning.

Extending Museum Literacy to the Science Classroom

Knowing how to use objects to extend the practice of science inquiry into the classroom can be empowering for the teacher because it allows the teacher to build a discourse of science around looking, observing, and generating hypotheses in a similar way to a practicing scientist. In addition, as the teacher has a choice in what they choose to display, the identity of the teacher (as a science teacher) also becomes central to how the visual culture is enacted in the classroom. For teachers who identify with being object-inquiry-based teachers, objects form a degree of symbolic capital. They are representative of the approaches to teaching that a teacher values—in this case, object/inquiry-based science teaching and learning. This object/inquiry-based approach is reflective of museum learning cultures in that objects are used to inspire and motivate scientific inquiry. In this respect, the museum and the object/inquiry-based classroom demonstrate a visual discourse of science and enable us to examine the relationships between looking, knowledge and power (Hooper-Greenhill, 2002) in relation to science teaching and learning. It also enables us to question the issue of power and agency within the museum field and how knowledge and enactment of the visual discourse in the science classroom begin to engage students in the practice of science.

In the expansivising space, emergent learning cultures form. The process could be described as a bricolage where the visual learning culture of the museum encounters the science classroom and develops new affordances for learning. In this space of bricolage, a teacher has the potential to enact an empowering culture of science teaching and learning, where they—the teacher and students—could take advantage of the free-choice (Falk, 2001) learning that characterizes learning with objects in a museum, while learning science content as required by the curriculum and standardized tests. This emergent, bricolaged learning culture values deep, critical looking as a way of coming to know, in addition to reading and writing, as the primary means of exchanging information and building meaning.

Objects as Representations

Objects are chosen for display in museums because they are exemplars of scientific concepts and phenomena. It could be for their uniqueness, aesthetics, value and/or role in scientific discovery; objects are selected to inspire visitors and engage their curiosity. Objects are imbued with meaning that people assign to them, and these meanings are socially and culturally bounded—even in the case of science, long positioned as a space of objectivity. Learning to do science in a museum entails learning about objects as representations of scientific facts and theories and learning the process behind the assignment of particular objects as exemplars of science.

Science objects ascribed as such can reify scientific concepts that may seem abstract. For example, one of my favorite displays in the Rose Center's Hall of the Universe is a case with various small Earth objects (a seashell, a piece of iron, and a beetle's carapace, among other things) with the following statement, "We Are Stardust. Every atom of oxygen in our lungs, of carbon in our muscles, of calcium in our bones, of iron in our blood—was created inside a star before Earth was born." These small objects seem random, but within the context of a museum display, they become objects of science imbued with the symbolic capital of being on display in a science museum. In this display, these particular objects become representations of the concept that all things on Earth are connected to each other and to the cosmos. They offer tangible evidence of conjectural ideas about the origins of planet Earth and the beings that inhabit and form it. Through viewing the object and reading the accompanying text, a learner can visualize their connections to things on Earth and beyond. Extending this to the classroom, having objects in the classroom can play a central role in science learning. Teachers could use objects to integrate the learning culture of museums associated with looking at and learning from objects and exhibits. This visual literacy encourages observing, which is an important practice in scientific research.

Objects and Identity

The Urban Advantage Lead Teachers used objects—both their personal objects and objects received or purchased for science—to re/create science displays in their classrooms. For the Lead Teachers, these collections of objects helped to construct their teaching identities, with collecting being a form of the production of self (Hooper-Greenhill, 2002). The display of science objects in the classroom are representations of the kinds of teachers the Lead Teachers were becoming: teachers who valued the object/inquiry-based approach as emphasized in Urban Advantage. In addition, the objects that teachers select to display are uniquely tied to their multiple identities, both in and out of the classroom.

Although museums take care to curate objects and display them, people interacting with the objects often ascribe their own meanings based on their own experiences with similar objects and/or any new thoughts or ideas that are inspired by their encounter with a new object. This is even more so when people with diverse histories and worldviews bring their perspectives to bear in their encounters with these objects. The story of Tomanowas, the Clackamas meteorite, highlights the historical and cultural conflict around some objects on display that are considered living and sacred to *their* people while inanimate and scientific to collectors, researchers and some visitors who view the object on display. It is important to consider the different ways that people encounter such objects and the implications of viewing and presenting them solely as objects of science.

The Urban Advantage Lead Teachers curated their displays in classrooms in ways that demonstrated their own lived meaning-making, identities, and values. The objects they selected to display and the ways that they encouraged students to engage with the objects all reflected the teachers' identities. The demonstration classrooms became one of the shared repertoires of the group, a conceptual and physical artifact that displayed who they were as teachers and the kinds of science learning they valued for their students.

Artifacts of Object/Inquiry-Based Science Teaching and Learning

"If you walked into an inquiry-based classroom, what would you expect to see?" This question started the initial Lead Teacher meeting. Lead Teachers brainstormed and developed a description of what an inquiry-based classroom would resemble. Central to the theme of the discussion was the display of science objects and resources that students could observe and manipulate. It was also crucial for the teachers to display student projects so students could see examples of science projects. This process enabled them to reify the object/inquiry-based classroom as they visualized/imagined and discussed their ideas. As a part of this brainstorm, they also visited a hands-on activity space in the museum where they experienced various observation stations, including preserved butterflies and mineral samples microscopes set up for viewing the fine details of objects and terrariums with living objects.

Animals and Magic

Display is a major form of pedagogy in the museum and became central to the classrooms of the Urban Advantage Lead Teachers. During visits to the Lead Teachers' classrooms, I noticed that several created cabinets filled with objects from their personal collections along with resources from Urban Advantage. Jake was a middle school teacher in Queens, and his classroom had several wooden cabinets with glass doors. One cabinet displayed human skull casts that he purchased with Urban Advantage funding, along with other bones, stones, balls, and plastic figurines. The adjacent cabinet had a collection of small toys and figurines from fast-food kiddie meals. Jake also had some small boxes filled with random objects, such as flat-end screws, wires, pen caps, popsicle sticks and rubber bands. He described these as "impromptu maker spaces" where students are encouraged to imagine and engineer tools and toys. Jake admitted that he likes to have a lot of stuff around because it "makes it interesting to the students;" this encouraged them to ask questions about his collections, and he viewed this as a way to foster inquiry by capturing their curiosities. This also allowed students a small view of his beyond-school identity.

Since Jake was a former zoo volunteer, one of the striking elements of his classroom was the variety of well-cared-for small animals, such as rabbits, gerbils, parakeets, and fish, each carefully labeled with the name of the animal and special instructions about handling animals.

When I visited Jake's classroom along with the Lead Teachers, I was brought back to my own middle school in Brooklyn, where my science teacher had an "Animal Lab." The Animal Lab teacher selected a group of students interested in learning about animals to eat lunch in the room once a week. While we were eating, he would then do a mini-lesson about one of the animals. When our meals were done, we *would spend the rest of the period just going around and looking at the animals or cleaning the cages. I remember him having a range of small animals, including mice, snakes, rabbits, guinea pigs, fish, birds, and lizards. They were one of the few spaces that helped to solidify my identity as someone who loved science.*

The Lead Teachers commented that Jake's classroom resembled a museum or small zoo; it was *"his* private zoo," as he referred to it. "I wish I had a teacher like you in junior high," one teacher remarked. Jake was immediately positioned as a good science teacher by his peers because of the quantity and breadth of objects in his classroom.

While the Lead Teachers worked to connect curricula and assessments to the resources and professional development of the cultural institutions, they also strived to enact the visuality of museums in the classroom. Amongst the Urban Advantage Lead Teachers, objects and classroom displays became identity markers of group affiliation.

The object-based demonstration classrooms became artifacts, visual confirmation of the integration of objects in science teaching practice. Amongst his peers, Jake's classroom became the exemplar. When we visited other teachers' rooms, they were quick to point out that their displays were not as elaborate or "good" as Jake's. However, although they did not aspire to create a classroom as sophisticated as Jake's, the Lead Teachers each aimed to create a classroom where objects were displayed and accessible to students as they were in Jake's space. Good science teaching, for the Lead Teachers, also meant having classrooms that inspired inquiry and creativity in their students. The demonstration classroom brought the museum to their students, which was important given the challenges of physically getting classes to the museum due to administrative policies and traveling logistics. These classrooms allowed students across the city to have access to learning with and through objects. They also served as models for other teachers in the set-up. While this did not make up for actual museum trips, objects in the classroom afforded students opportunities a similar enrichment in having close interactions with objects as a part of the ongoing science learning.

The objects in the classroom served as enrichment for students and were also salient expressions of teachers' identities and desires for their students. Jake's various objects displayed in the classroom represented a convergence of his multiple identities—his multiple memberships in different communities of practice—in his classroom. There were twirling rainbow wind catchers hanging from the fluorescent lights, and between the science objects were colorful toys and other seemingly random objects—random to me but carefully chosen by Jake to represent who he is as a science teacher and an individual. As a magician, he travels to the Midwest for his work during his vacation and brings back pieces of his sojourns to the classroom, like a giant cattle skull that he hung from a snake cage. Jake explained that the skull was used to demonstrate science concepts like anatomy and/or adaptations, but it can also be a window into a story about Jake's other identities. Many of Jake's objects had stories that he was able to share with students while using them to demonstrate science concepts. This sharing of self helps to build an element of trust, allowing students a window into Jake as a whole person; this could also encourage students to bring more of their whole selves into the classroom.

In many ways, Jake's classroom demonstrated the ways that a teacher's identity is a convergence in the classroom. Science teachers can bring different aspects of their roles into the classroom by practice and display. As the object-based learning culture of ISIs was being transformed into classroom practice, this also became an opportunity for the Lead Teachers to articulate their teaching philosophies and corresponding identity in their physical classroom. This was more evident in teachers who had ownership of their own classroom (versus having to share a room with other teachers and classes, as is more common in some schools).

Scientific Inquiry Zone

Pamela's classroom in the Bronx had a display shelf explicitly labeled "Scientific Inquiry Zone." The resources on the top shelf were received from one of the zoos, which included a snake's shed, a moth's cocoon, and other animal-related objects and specimens. Similar objects were used in an observation activity during the zoo professional development, where teachers were prompted to observe, generate questions, and aim to identify the objects in the jars. The rest of her collection included her own collection of rocks, books from various disciplines, magnifying glasses to facilitate observations, and, amusingly, a lunch bag with pictures of the solar system received during one of the museum workshops. I was surprised to see the bag there—that it was included in the scientific inquiry zone. She noted that she included it for aesthetics, "it was pretty," and that it was a visual representation of a topic covered in the curriculum.

Pamela commented about her classroom, "I have a lot of resources, grow tent, rocks, minerals, fossils, etc., … I would love some more open display cases because I ran out of room for all my stuff. I would like the kids to be able to look [at] and touch the resources." During an earlier meeting, Pamela mentioned, "I would like to have lots of stuff for the kids to touch. I want them to touch everything. I will put signs around that say, 'Please touch but don't break.'" Pamela was emphatic about having "stuff" and wanting students to interact with said "stuff." It was important to Pamela not only to be identified as an object/inquiry-based teacher but also to create these affordances for her students' engagement in science.

The display cabinets were examples of teachers creating new learning affordances that confirmed their identities as object/inquiry-based teachers. These displays also provided opportunities for their extra-teaching identities and knowledge to be related to their teaching identities. For example, Jake used his knowledge of animal-keeping and magic to create resources that his students could use to learn science. In a similar vein, Pamela's love of "stuff" translated into a classroom full of resources that afforded her students a learning culture of inquiry. The examples of Pamela and Jake's classrooms demonstrate how the museum's learning culture of the museum expansivised into the classroom and coalesced into affordances of object/inquiry-based science teaching and learning.

Cabinets of curiosities are viewed as precursors to museums. They were a way for wealthier people to display artifacts they acquired from travels and expeditions around the globe. It was interesting to see how these cabinets appeared in the Lead Teachers' science classrooms, where they not only represented the emergent learning culture that was being developed in the museum–school partnership but were also personal in how they represented the teacher who created them. As Pamela noted, these cabinets encourage students to look, touch, and ask questions and enable them to develop a mindset for doing science investigations.

Living Objects and Observations

All Urban Advantage schools received a grow tent from the botanical gardens to facilitate botanical inquiry in the classroom. In each of the Lead Teacher's classrooms I visited, the grow tent was set up with plants that were student projects and/or a variety of decorative plants received during workshops. As living and changing objects, the plants afforded long-term observations of changing phenomena and perhaps fostered patience in scientific observation.

In the different classrooms, the grow tents supported a range of object/inquiry-based projects. In one classroom, the grown tent contained a ginger plant (*Zingiber officinale*) and an *eddo* plant (*Colocasia esculenta*), the latter being a starch widely eaten in the Caribbean to demonstrate how some common foods from the market—cheap and widely available resources—could be used in the classroom. Another included a desert terrarium to help students learn about plant adaptations to different environments. These grow tents emphasized inquiry through curiosity through students' observation of living objects. A grow tent in a different classroom displayed student-controlled experiments, mirroring the process taught to the teacher during the professional development and labeled in a similar way that plants were labeled in the garden's greenhouse with the names of the plant (common and/or scientific) and date planted thus re/producing certain aspects of the practice of botany in the classroom. The grow tents afforded the culture of science inquiry, as done with living objects in the garden and in the classroom.

Visiting the Lead Teachers' science spaces revealed much of who they were and what they valued as science teachers. They experienced and learned different ways of knowing that sharpened the visual, aesthetic, and material understanding of the natural world. They were learning how to use those aspects of their personal and professional journeys to transform their physical classroom spaces. As a collective, they grew to identify as object/inquiry-based teachers who are experts in using objects in the classroom to support student learning and science investigation. The Lead Teachers used the new tools and objects, such as microscopes, globes, videos and new pedagogies around teaching with objects to create spaces that were bricolages of school and museum science. These resources became a part of who they are and revelations of their individual interests and passions about science that they shared with their students and in many cases their schools.

Demonstration Classrooms

In a Lead Teacher meeting where we discussed the issues and challenges of setting up demonstration classrooms, Elena mentioned she did not realize that her classroom was a demonstration classroom until district leaders visited her class [as an exemplar of science education in the district]. "The room was a mess, stuff was everywhere…they [the students] were busy working on their projects." The other Lead Teacher commented that her classroom *is* a demonstration classroom because the students were actively engaged in doing science. To the other teachers, the room, as Elena described it, sounded like a living classroom. At this point, the Lead Teachers had not visited Elena's classroom as a group; however, due to their shared visits and descriptions, they were able to imagine Elena's classroom as a demonstration classroom as their community defined it—students actively using the science resources and objects to do their science investigations. This collective imagination allowed the Lead Teachers to see themselves in new and different ways. The Lead Teachers' interactions and repertoire development enabled them to create collective and individual identities around their physical classrooms and ISE-related practices. The discussion about Elena's classroom allowed the teachers to articulate what defines them as a collective: what kind of teachers there are. Elena felt validated when her peers determined that she had an inquiry-based classroom; thus, in the Lead Teacher group, she was able to re-envision herself as a teacher with a demonstration inquiry-based classroom.

Assessment Mandates Constrain Affordances

One of the early theories of informal science learning—"free-choice" (Falk & Dierking, 1992)—describes the experience as self-paced and guided by the learner's interests. Guided inquiry has elements of free choice in that it is learner-centered, but it also includes prompts and/or scaffolding toward articulated learning goals and objectives (Miele & Adams, 2016). The Lead Teachers set up classrooms to foster inquiry along a continuum from free-choice/open-ended to guided inquiry. The displays and objects were designed to motivate students to ask questions and pursue their own interests and curiosities about the world. However, while the Lead Teachers afforded aspects of free-choice learning through their objects and displays, they also had to contend with the constraints of time and standardized test-related curricula, and this limited what the Lead Teachers were willing and able to do.

Although Jake had a variety of objects and resources representing the various disciplines of science, "I only allow my students to do animal behavior studies," as science projects, he continued, "I told my students that if they wanted to do something else, they were on their own...animal studies is what I know the most about." Jake talked about his time management skills, and since the Exit Projects are only worth 20 percent of the students' grades, he dedicates 20 percent of class time to working on the projects. That works out to one 45-minute period a week. The rest of his class time was dedicated to structured lessons that were more of the traditional chalk-and-talk approach rather than reflective of the ongoing inquiry that seemed evident in his classroom display.

Jake, like the other Lead Teachers, had a curriculum to cover that led up to the eighth-grade standardized test towards the end of the academic year. The curriculum encompassed a review of all the science content covered throughout middle school, leaving little time and space for students to pursue their interests in science. During her first year at Urban Advantage, Violet said she "made the mistake" of allowing her students to choose any topic for the Exit Project. "This made it really hard for me to help them all find information for their projects while trying to cover the curriculum," Violet commented. "Next year, I will narrow it down to one or two topics." However, the Lead Teachers desired and strived for student-centered practices, and standardized and mandated assessments created barriers to a full realization. The standards and assessments influences teachers' practices which then narrows the science learning opportunities for their students. As the Lead Teachers were in schools with intersections of racialized, immigrant, English language learner, low income, and diverse learner students, this provides insight into the ways that standards and assessments also serve to keep the most marginalized students excluded from meaningful and expansive STEM learning experiences.

Countering the constraints presented by the assessments, several of the Lead Teachers held after-school science clubs to maintain the space of inquiry for their students and allow them to enact their desire for object/inquiry-based science teaching. It was during the after-school session that students were able to work on their projects, and the teachers were able to fully enact inquiry-based science. Both Jake and Pamela discussed their after-school sessions: while students worked on their projects, they were also able to "play" with the other objects in the classroom as they explored different science topics. They described how students "hung out" in their classrooms and pursued a range of their interests. Even though the demonstration classrooms and teachers' dialogues indicated identities and desire towards object/inquiry-based teaching, the actual enactments fell short during regular teaching hours because of the impediments imposed by mandated assessments.

Wondering About Unused Resources

Urban Advantage provided teachers with resources to extend object-based science inquiry into the classroom. While many resources were used, some remained unused and, in some cases, stayed in their original shipping boxes. This presented a contradiction to the goal of using the resources to create object/inquiry-based classrooms.

There were many reasons given for the unused resources. The Lead Teachers cited that some of the resources did not arrive on time; for example rock samples arrived after they covered the rock cycle in class, therefore they would be used the next year. Others cited not knowing how to use specific resources. For example, some of the Lead Teachers did not set up the grow tents until after they attended a special session at the botanical garden, where they were given plants to start their collections. Some physical science resources remained unused because the teachers were not exposed to them during the workshop and did not have time to tinker with the resources to determine how best to use them. In one classroom that I visited, such resources were prominently displayed on the counter but still in their plastic shrink-wrap—not forgotten, but unused. The teacher clearly stated that she did not learn how to use them.

These observations confirmed that solely providing teachers with material resources would not lead to expanded science teaching and learning in the classroom. Providing resources alone does little to change learning culture or expand affordances, nor does simply learning about the resources and what they can or should be used for. Change happens when teachers are able to use resources (i.e., interacting with the objects during professional development) and then reflect on their enactments in the context of their classroom. Through collective reflections, teachers can begin to think of different ways to modify resources to meet specific goals and meet students' needs. Opportunities to use, reflect on and collectively modify images afford a broader range of enactment opportunities and spaces of agency for teachers to deepen and expand their practices. For example, in adapting an ISE learning culture practice of display, the rock samples would be kept out in a visible space and available to students to observe and interact with at times beyond the rock cycle lessons. This would allow students to see scientific phenomena as a part of an everyday activity rather than seasonal, like holiday decorations.

Identity in Practice

Objects embody feelings about our relations and are used to construct meaningful environments (Hooper-Greenhill, 2002). As evident in the Lead Teachers learning spaces, the objects were salient reflections of who they were and how they saw themselves as science teachers. As a collective, the Lead Teachers negotiated their identities around the creation of object/

inquiry-based demonstration classrooms. They defined and created structures that enabled the enactment of inquiry-based teaching and learning in the ISIs and schools. The demonstration classrooms were developed to show aspects of museum learning culture being merged with that of the classroom, along with the creation of an object/inquiry-based learning culture, thus creating a bricolage space for museum and classroom science teaching and learning culture. The display and use of the resources in the science classroom were cultural and symbolic capital for the teacher—cultural in her knowledge of the pedagogy of display and being able to effectively apply this pedagogy to setting up an inquiry-based classroom and symbolic in the association of lots of resources (or "stuff" as Pamela referred to it), with good science teaching and learning.

Interestingly, while learning with objects was a pedagogical practice across ISI settings, visual display was not explicitly taught as an educational practice. However, the Lead Teachers, immersed in learning cultures of display, somehow extended this culture into their classrooms. For the Lead Teachers, the object-based classroom inscribed upon them an identity of being a kind of teacher—an object/inquiry-based middle school science teacher who accesses, appropriates and takes ownership of a variety of resources to make science teaching and learning interesting for themselves and their students.

Teachers often decorate their classrooms to make them aesthetically engaging and to inspire student work. Different aspects of classroom visuals serve different purposes, some solely as decorations, whereas others as learning prompts and motivators. The displays I observed in the Lead Teachers classrooms were created for students to observe, ask questions, and collect scientific data. Carefully planned, deliberate display could foster inquiry and an investigative stance with objects (Hapsgood & Palinscar, 2004) and expand learning engagements for science in the classroom.

References

Bain, R., & Ellenbogen, K. (2002). Placing objects within disciplinary perspectives: Examples from history and science. In G. Paris (Ed.), *Perspectives on object-centered learning in museums* (pp. 153–170). Mahwah, NJ: Erlbaum Associates.

Bunch, L. (2021, April). Secretary Lonnie Bunch on "What Makes for a Great Museum Exhibition." *Smithsonian Magazine*. Retrieved from: https://www.smithsonianmag.com/smithsonian-institution/lonnie-bunch-great-museum-exhibition-180977158/

Falk, J. (2001). *Free-choice science education: How we learn science outside of school*. New York: Teacher's College Press.

Falk, J., & Dierking, L. (2000). *Learning from museums: Visitor experiences and the making of meaning.* Walnut Creek, CA: Altamira Press.

Hapgood, S., & Palinscar, A. (2002). Fostering and investigatory stance: Using text to mediate inquiry with museum objects. In S. Paris (Ed.), *Perspectives on object-centered learning in museums* (pp. 171–190). Mahwah, NJ: Erlbaum Associates.

Hooper-Greenhill, E. (2000). *Museums and the interpretation of visual culture.* New York: Routledge.

Miele, E. A., & Adams, J. D. (2016). Guided-choice learning in out-of-school environments. *Science Scope, 39*(6), 52.

Rowe, S. (2002). The role of objects in active, distributed meaning-making. In G. Paris (Ed.), *Perspectives on object-centered learning in museums* (pp. 19–36). Mahwah, NJ: Erlbaum Associates.

· 5 ·

EXPANDED AGENCY THROUGH MODIFYING AND CREATING AFFORDANCES

The "Bad School"

In the early 1980s, Central Brooklyn Middle School (CBMS) had the reputation of being a "bad school." Situated in the heart of a predominantly Afro-Caribbean community, it was our zoned school, and parents who were aware of the reputation resisted enrolling their kids there. However, due to the difficulty of enrolling in schools outside of the neighborhood, my brother ended up attending for one year—which was about all he could handle. My brother told stories of students constantly disrupting the class and cussing at the teachers, all creating challenges for meaningful learning in the school. I visited the school once with my parents during parent-teacher conference night, and while I don't remember much, I remember the halls and classrooms being drab and dirty in contrast to the clean, bright halls of the gifted and talented and predominantly white junior high I was bussed to in a white different community.

Several decades later, I visited Violet to support her in recruiting more science teachers from her school for Urban Advantage. Because the school was situated in the community that I had long called home, I had a personal investment in supporting this school in Urban Advantage. I was excited that Violet, as a teacher from this school, chose to be a Lead Teacher for Urban Advantage, and I was also curious to see how school changed.

When I first entered the school, I noted a different building. The halls were clean and brightly lit, and I saw acrylic self-portraits of students displayed in the foyer. The 1st floor halls were clean and quiet, with student work and art displays on the walls. The English Language Arts (ELA) assessment rubrics and related student work were also displayed prominently. I looked for evidence of science on these bulletin boards but saw none.

The Existing Learning Culture for Science

The science teachers in Urban Advantage had desires to engage in science education opportunities for themselves and for their students. While some of the teachers came to the initiative because they were sent by their building or district administrator, most volunteered because they sought out these kinds of teacher learning opportunities to support the science teaching identities they saw for themselves. The Lead Teachers were also a self-selected group. The professional development resonated with them, making them want to play a more central role in the initiative. They valued the cultural institutions for the resources they could provide for the city's middle schoolers. It would make it more relevant and engaging for students and more interesting and intellectually challenging for teachers.

One of the goals of Urban Advantage was to establish sustainable practices for teachers and students using cultural institutions to teach and learn science. The demo classrooms' development helped show the integration of museum-based teaching practices in the classroom. These classrooms also provided examples for other teachers to observe object/inquiry-based classrooms and imagine what these spaces would look like in their own schools.

As described in Chapter Four, the development of these classrooms demonstrated the ways that the Lead Teachers took ownership of the learning culture that participated in the museum and created learning spaces that were reflective of the learning cultures of the classroom, the museum as well as their own identities and lived experiences. The Lead Teachers collectively developed practices and science teaching identities around the notion of object/inquiry-based pedagogy inside and outside the classroom.

In this chapter, I turn the lens to one of the Lead Teachers, Violet, in order to describe the process of her expanded restorative agency as manifested in the creation of her demo classroom and in her interactions with her school-based peers. Here, I will discuss the transformative process in a school and classroom that began with a teacher who values using science resources and identifies with using museum-based resources to teach.

The Bathroom Pass

While Violet was presenting a PowerPoint lesson demonstrating how students could organize their research, a student raised his hand and asked for a bathroom pass. "Write one out, and I will sign it," Violet responded without missing a beat in her lesson. He asked a classmate for a piece of paper, and she ripped one from her Urban Advantage notebook. When he presented Violet with the pass, she abruptly stopped her lesson. "That is disrespectful! Urban Advantage gave us these resources. Are you going to use that paper for a bathroom pass with Ms. Adams standing here?" Embarrassed, the student tried to explain that another student had given it to him (in true middle school fashion, that student was giggling because he got in trouble). "You will apologize to Ms. Adams right now!" The student apologized and sulked to his seat, and Violet resumed her lesson. To Violet, classroom resources and notebooks were imbued with special meaning and associated practices of object/inquiry-based science and connection to the Urban Advantage community. To the student, it was just a notebook to be used to meet immediate needs, in this case for access to the bathroom; however, to Violet, the notebook was a valuable science resource and symbolic of the ways that Violet's teaching practices evolved over the year, as a Urban Advantage Lead Teacher.

Keeping science notebooks was one of the practices of scientific research that was emphasized in Urban Advantage professional development. Teachers saw examples of scientists' notebooks and were encouraged to keep their own during the professional development to document their own learning and investigations. For Violet, the notebooks were an important connection between the science research in the museum and her students' learning in the classroom and came to hold a central place in her classroom practice.

From the beginning of my interactions with Violet, she always emphasized the "wonderful resources" that Urban Advantage provided to her classroom, including the vouchers that allowed her and her students free access to cultural institutions. She mentioned that her school had been involved in several parachute initiatives since she had been there for about three years. "They charge a lot of money and come in a couple of times for professional development, but you never see them again. Urban Advantage provides professional development *and* resources." Resources provide anchors for learning cultures; they form the basis of affordances and help "create a rhythm of engagement, imagination and alignment" (Wenger, 1998, p. 250) with the emergent goals of an expansivising learning culture. Having access to and knowledge of using a body of resources is also an affordance for a group to become a community and align itself with certain goals and practices around using those resources. Science resources—teacher education, access to cultural institutions and equipment—were central to the initiative, and the Urban Advantage learning community developed practices around integrating the resources and the schema (i.e., science content knowledge and ideologies around museum-based learning) into their teaching practice. This was especially evident in the Lead Teacher group, which Violet was a member of. To Violet, the Urban Advantage resources demonstrated a true commitment to the school; in turn, she was committed to using the resources to enact in her classroom the inquiry-based pedagogy that she learned in Urban Advantage, which shaped her identity as an object/inquiry-based science teacher.

Neglect of Science

Central Brooklyn Middle School was designated as a "high needs" school. High-needs schools have large numbers of students characterized as "diverse learners;" Black and/or Latinx students and/or culturally and linguistically diverse and/or qualified for the federal free/reduced lunch program (Avalos, Perez, and Thorrington, 2020) and/or neurologically, cognitively and/or physically diverse; in essence students who have been historically underserved in public education. Students in these designated "high needs" schools often do not meet mandated levels on the standardized English Language Arts (literacy) and math assessments. According to Violet and other teachers I met from the school, much of the instructional focus was on the literacy and math assessments, leaving science a low priority in the school. This was manifested in observing the minimally used [dusty] lab with broken chairs, excessively scratched tables, and busted archaic random equipment, which were a testament to the long neglect of science. The lab had the appearance of an old, abandoned factory in an economically failed mill town. At the time, it was the only "usable" lab space in the school, "There are two other labs, but I have never seen them, and they are not used," Violet pointed out.

"Most of the time, this lab is empty," Violet commented as we looked at the small whiteboard in the front of the room with a schedule grid that included the teachers' names and the periods they were scheduled to use the lab. The lab had eight blacktopped lab tables, and each table could accommodate four to six students comfortably. It is a good set-up for group work, so Violet used the lab for her students to do group work on their Exit Projects.

Besides Violet's class, I only saw the lab actively in use once. This was during the eighth-grade science practical exams. The lab was set up in stations weeks before so the students could practice the assessment skills and sit for the exam. This was in early March, and during subsequent visits up until late May, the practical exam was still in place, although the implementation was complete.

Comparing Science Classrooms

Violet noticed the disparities in schools when she participated in the Lead Teacher intervisitations. Upon seeing the science learning spaces in a middle class mostly white and Asian school, she remarked, "Wow, now I can see where they put the resources." The school had a brand-new science lab equipped with hood and safety equipment donated by a local politician. She also saw that teachers' science classrooms were filled with a large variety and number of science resources. Violet's lab, in contrast, only had a few plastic cylinders and obsolete computers pushed in a corner. "You see, they do not put the resources here [in her school] to keep the community down. Uneducated children become uneducated adults that never leave the community." In referring to the collective "they," Violet noted the systemic issues that inequitably influence the education of Black, Latinx and lower-income students across the city. Schools in such communities are often under-resourced and deemphasize science in favor of math and literacy mandates.

When I first visited Violet's school to begin my case study, one of the first things I noticed was the student artwork that decorated the office and the halls. In the office, there were popsicle-stick houses painted in the pinks and blues of houses in the Caribbean. Hanging in the halls were 3-D collages painted in the colors of Caribbean flags. As I walked through the halls towards Violet's classroom, I saw some students with flag bracelets and bandanas, and I heard the accents and idioms familiar to me. When I visited her Assistant Principal's office, I noticed his Trinidadian flag and a picture of a beach scene facing his desk, perhaps a reminder of a quiet spot "back home."

Being situated in an Afro-Caribbean community, the school reflects an unequal system where "this inequality affects their belief in the value of an education." (Bobb & Clarke, 2001, p. 224). Caribbean immigrants in these communities learn that Black schools are less valued than white schools and come to internalize the notion that the system does not care about Black students. When Violet saw the well-equipped, bright lab, she immediately made the association between the better lab and the non-Black racial demographics of the community where the school is located.

Violet's school demonstrated this inequity through the physical manifestation of unused and/or outdated science classrooms. This also confirms that science learning is not important for students in high-needs schools and reinforces the deficit-oriented idea that students in such schools are incapable of learning science. Further, this supports the notion that providing science to these schools and students is an altruistic act of service rather than an educational right and runs the risk of repeating early museum and philanthropic ideologies of "elevating the less fortunate." However, there are teachers like Violet who resist these ideas and view science as a means for an equitable and just education. Their activism lies in their sense of restorative agency and the conviction that rich and meaningful science learning is imperative for their students, not just the ones in the more affluent and historically white communities.

For Violet, transforming her classroom was one way of achieving science equity for her students. While the lab was in a dormant state, Violet's classroom had evolved into the place that she dreamed of, with working stations, displays, and student Exit Projects placed around the perimeter of the room. A door separated Violet's classroom from the lab, and that door also separated two contrasting views of science in the school. On Violet's side of the door, science is a practice, a skill or a way of thinking that could be used in life. "I want the students to see science as everyday life. They could use scientific thinking to solve everyday problems," Violet said when talking about her view of teaching science. The resources are always present, reflecting Violet's view of the ever-presence of science. The contrasting view in the lab is a hidden or forgotten science only to be revived when it is a priority for assessments. In this view, it is not a central practice of living or thinking but a subject that is only as important as the mandated assessments make it. The lab only became an active lab in response to an exam.

In CBMS, the teachers travel to the students rather than the other way around, as in many middle schools; therefore, the students spend most of their day in their homeroom classroom. Violet's science classroom was also a homeroom and was not used solely for science institution. As Violet explained, "Normally, the students have science in their classrooms. Carts are provided for teachers to take any materials that would enhance the lesson." This contrasts with the other Lead Teachers who had designated science classrooms where science objects were always on display and accessible to students and teachers. Violet mentioned several times how traveling with resources and teaching science in a room not set up for science teaching and learning was challenging.

Science did not have a space in Violet's school, neither as a physical classroom nor as a curriculum. Violet labored to extend the expansiving space of the museum–school learning culture that she both participated in and helped create in her school. She pushed this learning culture around her science cart and the students' homerooms. The objects on her cart seemed diminished by the overwhelming array of literacy and math-related visuals and artifacts in the room. Even the students' desks had neon-colored essay writing rubrics taped in the upper left-hand corners. The science traveled with Violet, present when she was and absent when she left.

Envisioning a Science Space

Violet had a vision of science becoming central to the practice of education at CBMS, "My dream would be to see every student at [CBMS] as a scientist." For her students and the school, Violet wanted her homeroom to be a space that would be available for students to learn and engage with the science resources.

"This is where I would like to put the resources." Violet opened some empty overhead cabinets above the counter with plastic baskets of books, videos, and other science-curricular materials. It was Violet's desire to create "a place where students could come and do projects" and to display them once they were done. Violet's classroom, which was also a sixth-grade homeroom, had evidence of her efforts to bring object/inquiry-based science to the school. Amidst the literacy posters and rubrics, Urban Advantage science resources were waiting for their chance to be prominently displayed and moved to a central space in the instructional affordances of the classroom. The grow tent was in the back of the room, set up with plants Violet had received from the botanical garden during a recent Lead Teacher meeting. On the right side of the room was a little station set up with two Petri dishes—one containing a dead beetle and the other a dried maggot. She had the students use the dissecting microscope to observe these specimens and draw them in their notebooks. On the same table were two jars—one with a brownish fluid and the other with water plants (also from the botanical garden)—set up to be an observation station. It appeared that the Urban Advantage resources were the only science resources in the room. Violet was proud of her classroom and the resources that she had displayed. She also knew that she wanted this room to be *the* science room and available to all teachers and students in the school.

Violet's homeroom shifted to one that demonstrated her Urban Advantage Lead Teacher identity and the expansiving space that she co-created with her peers. Her classroom evolved to reflect a space where inquiry-based science happens. "I wanted the students to have something to observe and write in their notebooks." Violet wanted to start the term with the habits of observing and recording, so she used the dissecting microscopes that she received during the summer Lead Teacher workshops to set up mini observation stations in her classroom. Violet wanted to start the year by implementing key scientific inquiry practices in her classroom. Violet used the microscopes and notebooks to change the affordances available for student science learning. They were able to engage in inquiry/object-based learning as emphasized in Urban Advantage pedagogy.

Claiming a Space for Science

"We will have science in this school," Violet proclaimed during one of my visits. With her ongoing participation as a Lead Teacher, Violet became increasingly resolved in her vision to have a science classroom and expanded science learning affordances for all the students and teachers at CBMS.

> "My interpretation was to use this classroom to teach students. Teachers could take their kids there for special lessons. It was my suggestion. We did not have a room, especially for science, so I suggested that the principal use it as a model classroom. When I brought my students there for a lesson, [my principal asked] 'Why are you bringing students here? It is supposed to be a model classroom for Professional Development [for the teachers].' I told him if it will be a resource room [for teachers], then we will need another room to be a demo room [for students]."
>
> Violet wanted a space for students to practice science; this seemed to conflict with her principal's idea of having a resource room for teachers. Violet wanted the room available so that other teachers could come and use it *with their students* when it was available. In keeping with her vision of having all students at CBMS School as scientists, she wanted her room to become where students and their teachers could come and *be* scientists.
>
> Violet's new classroom was formerly occupied by a literacy homeroom. This was evidenced by the array of literacy rubrics and slogans that decorated the room. There was little room for science. Violet was given the room after the literacy teacher was promoted and her classes dissolved. Like a realtor pointing out the potential of an old building, she pointed out all the different stations that she would set up—one for Earth science, another for life science and one for physical science. In a corner

near the windows, there was a dusty bookshelf that would be cleaned and used for reference books and videos. With her teacher's choice money, she hoped to purchase a TV and VCR for the room.

Violet's participation in the Lead Teacher group afforded her access to resources that gave her the impetus to transform science education in her school. With Urban Advantage, Violet had access to resources, both material and human, in the form of science equipment—my relatively frequent presence in the school afforded her a resource and a connection with the museum. She often introduced me to her colleagues and students as the one who brought Urban Advantage to the school. In addition, as a Lead Teacher, she became a member of a community of teachers whose focus was to build science education capacity in their schools. Violet agreed that visiting the other teachers' classrooms during Lead Teacher meetings "gave her ideas and encouragement to work on the demo classroom." Her finished classroom was the physical representation of her participation in the Lead Teacher community of practice, as it contained certain aspects of the intervisitations and her work with the museum. Through her association with me and the Lead Teacher group and her knowledge of setting up a science classroom, Violet was empowered to advocate for her classroom to become the room dedicated to science in her school.

"There was some more discussion about the demo classroom. It was agreed that depending on how it was set up, some students would be allowed to use this class [room] for particular lessons. The principal was also trying to obtain some computers for the room," Violet's sense of restorative agency catalyzed access to this space for science, "It is left up to me now to have the room ready for use. I have used up two Saturdays to start fixing it. Seems as though I would have to do this for a few more weeks because it is difficult to work on it during the regular school week; however, it has already shown signs of development," Violet reported to me in an email. When I visited the school shortly after this email, I saw the signs of the room coming together. Posters depicting the Solar System and a chart listing the steps of the scientific method replaced the literacy posters, and literacy essays were replaced by chart paper listing students' inquiry questions gathered from a recent museum trip. Many resources were moved out of the closet and onto the counters around the room. For example, using milk crates, she set up an ad hoc display case and included objects from her own personal collection as well as some that she received from one of the zoos. Although it was a modest display, having this in her classroom marked

Violet as a member of the Lead Teachers community and identified her as an object/inquiry-based teacher.

For Violet, the science equipment represented resources and confirmed her identity as a Leader in Urban Advantage and the museum. The resources were of great value to her because of how she can use and modify them to create expanded affordances for science learning in her school. The classroom demonstrated her identity as an object/inquiry-based teacher who practices scientific objects and visuals to support her students' engagement with science. For Violet, affordances for meaningful science learning were a matter of justice for her students and her community.

Extending the Science Resources to the School

"There is a need for science teachers to get together and make science work for the school," Violet told me in an email exchange. Violet held a meeting for the science teachers in her demonstration classroom, where she discussed Urban Advantage and demonstrated some of the resources that were provided for the school. Violet began the meeting by describing Urban Advantage as "a great opportunity, and I think we need to take full advantage of it." She gave each teacher a carefully prepared folder with informational documents about Urban Advantage and two inquiry-based worksheets, one she received during a Urban Advantage workshop. While she reviewed these documents, the teachers quietly followed along, referring to the same documents in their folders. Violet invited me to this meeting to support recruiting more teachers in her school for Urban Advantage. She introduced me as "the one who brought Urban Advantage to the school," she mentioned on several occasions that it was an honor to have me in the school.

"Now I must show you some of the things that we have from Urban Advantage, and they are not [only] for Violet Williams; they are for the school; these are here for your use." Violet moved to the part of the meeting where she introduced the teachers to the Urban Advantage resources. It was clear that while she had the resources, she wanted the other teachers to use them. This goes back to her personal mission of bringing science back to the school. Violet's acquisition of the resources and her ability to modify them for classroom use put her in the position of extending the expansivising learning culture into her school. As Violet continues to take out the resources, the buzz around the room increases, and much excitement is generated by the idea of having access to the assortment of science resources presented.

Violet's enthusiasm for the initiative sparked enthusiasm amongst her peers, prompting 4 of her colleagues to sign up and become an official part of Urban Advantage, which meant having access to professional development, additional resources for the school and classroom and field trips for families and students.

In schools with meager resources for science teaching, there are often tensions between teachers who are viewed as "having" the resources and those who do not. Although the Urban Advantage resources were meant for schoolwide use, the Lead Teachers often mentioned not wanting to be viewed as hoarding resources or being seen as "keeping the resources for themselves" or "the one who gets everything." In urban public schools, there is often inequitable access to science resources across the city and even within schools; therefore, there is the potential for conflict between the "haves" and the "have-nots." With effective science teaching being resource-dependent, the Lead Teachers mentioned that while they did not want to be shunned by their peers, they wanted to get as many resources as possible for their classrooms. They wanted nice demonstration classrooms and had the resources available to confirm their identities as object/inquiry-based teachers. While they kept certain resources for their classrooms, they were also mindful of getting extras to share with their peers wherever they could.

Violet also experienced this tension in her school. For example, when an issue arose about the sixth-grade teachers participating in Urban Advantage, Violet was more than happy to have me explain to one of the teachers why they were not able to receive the full resources, "I do not want them to think that I am keeping anything from them," Violet mentioned. She was happy to have the resources for the school; however, she seemed conflicted about having the power that came with having the resources. Also adding to Violet's discomfort was that since she was relatively new to the school, she probably wondered what other teachers would think about her getting all of the "stuff"— if there would be feelings of resentment or jealousy. This would contradict her goal of building solidarity amongst the science teachers for science at Central Brooklyn Middle School.

Expanding Science Within the Literacy Space

About six months into the first Lead Teacher year, I visited Violet's school with two educators from one of the Botanical Garden partners. Brenda and Linda were interested because of the number of Urban Advantage teachers Violet recruited from her school. Several of the teachers participated in workshops at the Garden as it is quite close to the school. When we entered Violet's classroom, which was now the science resource room, we were pleasantly greeted by the smell of Chinese takeaway and an array of goodies that Violet had prepared for us, her guests. Alongside the prominently placed science resources, student projects were also displayed around the room to show the work that students have done with the Urban Advantage resources. While we waited for the meeting to begin, one of the sixth-grade teachers approached us, "Ms. Violet said that you would like to see our classrooms."

We followed her downstairs to her room. When Brenda, Linda and I entered the classroom, as I expected, the space was dominated by literacy and math posters and student work. She led us straight to a small rack of test tubes propped on a windowsill, each filled with a clear liquid that looked like water from across the room. Proudly, she pointed out to us that these were the DNA extractions of various plants that she did with her students. Posted near the widow was a small poster of a plant cell. Amidst the literacy in this teacher's classroom, the display of DNA extractions in this teacher's classroom was evidence of the beginning of the expansion of science into this literacy space.

With Violet's involvement in Urban Advantage. the school had an emergent school-wide science learning culture. Her enthusiasm around the resources and efforts to create a science resource room for the school seemed to create catalytic energy in the school, even among teachers who were not science teachers. Although Urban Advantage was earmarked for seventh and eighth-grade students, the sixth-grade teachers became interested because of the resources the initiative brought to their school. The sixth-grade teachers even brought their students to the end-of-year city-wide science fair at the museum so they could see what would be required of them in the later grades.

The Established Room

"The science classroom will be extended to all science teachers and science students. It is expected that we will all benefit from it," Violet declared in one of her emails to me. The classroom finally established its place as central to science teaching and learning at CBMS. During one of my last visits to the school that year, I got to see the completed classroom. Since it was late in the term and some of the resources had been packed away for the summer, I did not get to see the room as fully set up, but there was evidence that it was an inquiry-based science classroom.

Student projects decorated the room's perimeter, and there were several student artifacts—two terrariums demonstrating global warming, beakers filled with crystals from a crystal growth experiment and ginger plants placed around the room alongside their corresponding projects. All of the projects demonstrated elements of Violet's teacher learning and the expansivising space between two learning cultures. One of the projects pictured utilized fish tanks purchased through Urban Advantage to explore "global warming," one of the topics covered in the Urban Advantage professional development. These students became very interested in global warming after learning about the topic from Violet and viewing exhibits addressing global warming in the museum. They used the fish tank to demonstrate global warming—their visual display of a science topic.

Another project utilized the grow tent and project idea from the botanical garden. The student grew three common plants—an avocado, ginger, and tomato—to see how they grew compared to one another. In the garden, the professional development included growing plants from foods commonly found in the market. Students using science to learn about their world was Violet's vision of science at CBMS, and these projects were evidence of Violet's desires manifesting themselves.

The remainder of the projects displayed varied topics, from mosquitoes to growing crystals. Violet allowed the students to explore whatever questions they had. It was important to her to foster their curiosity and personal interests, even though she later realized that managing such a wide range of topics was difficult for her. However, she used all the resources she had at her disposal to help her students pursue their diverse interests in science.

As the eighth-grade assessment covered various science topics, the displayed student projects demonstrated how learning in various contexts supported Violet's teaching and her students' learning across the science curriculum. In the botanical gardens, she learned how to do life science studies with plants and at the Natural History Museum, she learned how to do field investigations to study the Earth's processes. What Violet learned became a part of her classroom—she created her bricolage of science by moving in between and learning in different learning cultures and transforming science in her classroom and her school. In between the student projects were some more recently arrived Urban Advantage resources that were still in boxes on the counter. While there was ample evidence that Violet had expanded the partnership into the school, the culture of assessments still dominated; Violet and her colleagues were using the space to grade the recent eighth-grade exams.

Transformed Spaces and Emerging Genius

Jamal was a tall, slender student, the tallest among his peers. He was also a Caribbean middle school "bad boy," one who was purposefully defiant and disinterested in school and especially science. The other students looked up to him because he was perceived as being gangsta-cool. When I visited Violet's class, he was one of the students that she pointed out as challenging. He was frequently absent, and when he did come to class, he was often disruptive. He was a known entity to the school security guards, and as a ringleader, he set the tone. Some of the other students, the boys in particular, followed along.

When Violet started to bring some of the Urban Advantage affordances into the classroom, she noticed changes in Jamal. Not only was he showing up more consistently, but he was also less disruptive and less attentive. More importantly, the objects in the classroom piqued his curiosity, leading him to observe and pose questions about the objects. The other students took note, and because Jamal was cool and wanted to be cool, they became more deliberate in their positive engagements.

In critically oriented research, we often talk about being a disruptor of the status quo. Jamal was a disruptor of the silent, irrelevant science classroom and became an enabler of science engagement for himself and his peers. He and his classmates were in the program's first year; Violet did some mini-investigations to familiarize them with the learning approaches and process. Jamil was interested in what Violet shared with them from the botanical garden, where they grew plants from fruits, seeds, and tubers found in the local Caribbean markets. He led his group in growing bell peppers. They yielded a very lovely red one that they then wanted to sell, which Violet forbade.

Jamal took his interest in science home. Violet received a call from the local precinct because apparently Jamal was experimenting outside of his home with baking soda and set off a minor explosion. The police officer wanted to confirm that Jamal was actually doing science homework. While it was not something that Violet assigned, Violet confirmed with the cop and later discussed what kinds of experiments would be safe for him to do at home, especially given the relationship between the Black community and law enforcement. As an aside, Jamal's life was probably spared because the officer had a positive relationship with the school and used his sense of community and humanity to speak to the teacher rather than take unnecessary action against Jamal.

In the eighth grade, students were required to do the Exit Projects. This year, Jamal became curious about space science. Violet took her class to the Space Show and brought some of the digital resources back to the classroom so that the students could view them on their own. Jamal was fascinated by Jupiter, particularly the planet's size and spinning speed. He decided to pursue his independent project on Jupiter and started to do his own research into more information about Jupiter. During his inquiry, a question about Jupiter's rotation emerged, and Jamal could not determine the answer. He reached out to a NASA scientist. It turns out that not only did the scientist not have any information about Jamal's inquiry, but he had not thought or heard of the question himself! Jamal became a hero amongst his peers for stumping the scientist. The student who was formerly positioned as a "bad bwoy" was now a "wicked scientist."

Expansivising the Learning Culture of the School

Violet learned the affordances of the museum and, within the Lead Teachers group, modified meanings and enactments of museum learning culture in their classrooms. This expanded her agency as a science teacher, which she then used to transform science teaching and learning in her school. More so, her restorative agency in expanding the affordances for science learning for her students. She used the resources and schema generated in the expansivising space to create affordances for an object/inquiry-based science classroom. She also became an affordance for other teachers in her school. Her affiliation with the Lead Teachers enabled her to identify with a community of educators with similar goals and developing practices. The opportunity to visit the other Lead Teachers' demonstration classrooms confirmed her group affiliation as an object/inquiry-based teacher. It enabled her to envision her classroom as an active inquiry-based learning space. She then used the Urban Advantage resources to create a demonstration classroom and use her Urban Advantage affordances to extend science into literacy-dominated spaces. In doing so, she was able to extend her identity as a scientist, or "a curious person" as she described herself to her students, by then creating the spaces and affordances for her own students to be their own curious people and have the resources available for them to follow their creative-science paths.

The learning culture that Violet created/participated in within expansivising space extended into her school, leading to school-wide changes in science education. In CBMS, Violet served as an affordance for other teachers in sharing her knowledge of science, the Urban Advantage physical resources of science, and in-school professional development on how to use the resources with students. Her ultimate goal was to create a space where teachers could come and plan their lessons and gather the equipment they would need to demonstrate science concepts. Violet was well on her way to realizing this goal and was looking forward to starting the following school year with a well-equipped demonstration classroom.

Violet's Prequel

My first visits to CBMS saw a building lacking in science resources. Violet showed me the vacant science lab and her "science" classroom, which was dominated by literacy and math mandates. It was when I had the opportunity to discuss Violet's school with her during a summer museum program unrelated to Urban Advantage that I found out that Violet thought that her school was lacking in resources, too, "that's because I did not know about the resources. It wasn't until Urban Advantage start[ed] pumping materials into the school and I start[ed] sharing with teachers, then is when they let me know that 'look, this is what we have.'" The teachers who were in CBMS long before Violet revealed to her the resources that were tucked away in some of the classrooms and closets. Violet previously thought that the only resources that were available to her were the ones that she received from Urban Advantage:

> Now, I am getting excited that I'm getting materials to work with; I don't want my classes alone to use these materials. [I] share[d] it with the other classes so that we could get the science going. Then, I was told that this school has tons and tons of materials. Take, for instance, the rock samples that we got from Urban Advantage. I carried the rock samples, and I showed them to them. I said, "Look what we have!" Then the teachers told me, "We have a whole room full of rock[s]." When I tell you rock[s], we had thick rocks about this size (gestured with hands)— 5 lbs or more.

I asked Violet if the other teachers used the resources, and she said that they did not, "In fact, science teachers [here] are seen as 'teachers who teach on their seat and not on their feet.'" The teachers that were there before Violet were working within a school culture that "played down" science, as

Violet described, so the teachers were not encouraged to go beyond covering the curriculum,

> There is one teacher there especially who said he was very enthusiastic when he arrived at the school five years ago. [He had] tons and tons of ideas, but everything he put forward was put down. It was put down by the administration and then by the students.

The administration did not regard science instruction as important and thus contributed to the minimal effort in science teaching and learning. This also conveyed to students that science was not important since it was not something that was accompanied by a standardized assessment.

Despite this history and challenge, Violet maintained a positive outlook on science teaching and learning; she mentioned that she has been doing whatever she can to make the lessons exciting because "if the lessons are exciting, the students will want to learn." She extends her excitement towards getting teachers to use the resources, "The materials are there, but somebody needs to get somebody excited. And I have a feeling that might be my job (laughs)." She believes that the science classroom that she has set up will be instrumental in bringing the teachers around,

> I know that these are teachers who can work, but the will is not there. That is why I am hoping that this science classroom that I am given permission to set up and operate this school year and that I could fix it in such a way that the administration allows teachers to use this room for at least preparation. And if we have a special lesson, say, for instance, if we are looking at the use of the pulleys, we could set up pulleys in the room and have students come to the room rather than have to dismantle them and take them to the classrooms every time. I am hoping that we could do things like that so that teachers could be encouraged. So, for this new school year, I have not been given a homeroom. I'm given the responsibility to have that room ready. I think we have made strides in that area.

Through co-learning with the Lead Teachers, Violet became *the* critical agent for science teaching and learning at CBMS. She was able to bring the material resources of science into her school and allowed the emergence of long-dormant science resources into active spaces. In the wake of Violet's expansivising, space waves of new affordances for science were created and expanded opportunities for her students to imagine themselves as science people.

References

Avalos, M. A., Perez, X., & Thorrington, V. (2020). Comparing secondary English teachers' ideal and actual writing practices for diverse learners: Constrained professionalism in figured worlds of high-stakes testing. *Reading and Writing Quarterly, 36*(3), 225–242.

Gee, J. P. (2000–2001). Identity as an analytic lens for research in education. *Review of Research in Education, 25,* 99–125.

Wenger, E. (1998). *Communities of practice: Learning meaning and identity.* Cambridge, UK: Cambridge University Press.

Initial Meeting

There was a good turnout this evening. There were three familiar faces and many new ones. After going over the particulars of the project, I asked them to reflect on how they think ILE learning influenced their teaching practice. They spoke about connecting experiences to the classroom by reproducing the activities they learned in the course and workshops; however, what stood out for me was their spirit of advocacy for science in their schools. They spoke about the marginalization of science compared to math and ELA and how it is often given low priority by their administrators (lack of administrative support for science). However, they work in both blatant and subtle ways to advocate for science for their students. One teacher noted that her administrator mentioned that science is only about learning facts, diminishing the importance of science, to which she passionately replied, "It is much more than that—it is about getting students engaged in doing science both inside and outside of the classroom." They spoke about science in terms of curiosity and motivation.

When we moved on to a discussion about their informal science learning, they described how they have been enacting informal science in their classrooms; for example, how they used the resources that they received from the museum "to enhance science engagement", co-investigating Daphnia with their students, using student questions as a resource for their own teacher learning, and engaging students in self-directed learning. All of these things were offered well before we began to define what informal science meant to us as a collective. One teacher questioned, "How do I keep it structured while making it informal?" Teachers valued the emergence of informal science learning but were also concerned about meeting the curricular requirements in the classroom that had both content and temporal constraints. It will be interesting to see how teachers deal with the affordances of ISE while navigating the very real constraints that emerge in the classroom.

This was a part of my reflections/memos from my first meeting with the ILETES teachers, the Collaborative Teacher Inquiry for Informal Science Learning and Science Teaching in Urban Classrooms (hereafter the Collaborative). They were recruited to participate in the research because they took preservice courses that integrated ISE and/or participated in extended professional development, like Urban Advantage, where accessing the affordances of ISE was central to the pedagogical goals. The goal of this meeting was to provide an overview of the research project and to begin to learn about their initial ideas about ISE learning. I knew several of the teachers from prior interactions—I taught them in preservice learning courses, and several of them were associates in the National Science Foundation's Robert Noyce Teacher Scholarship program[1].

However, as a group, we were getting to know each other in a more informal, collaborative way. I designed the meetings around cogenerative dialogues, a methodological approach where participants discuss shared experiences and issues with the goal of catalyzing change (Martin, 2006; Tobin, 2006). With the notion that people make meanings, shape identities, and identify modes of action through discourse (Luke, 1995), I aimed for the meetings to be participatory and dialogic and for the teachers to think about it as a part of their professional learning. I wanted the Collaborative to self-guide their learning as a collective of teachers, so I prompted them to think about questions they have about their classroom, students and learning and bring these to the group meetings to shape the dialogue. Although they received a stipend for their participation, I prioritized their learning and navigations of teaching science in schools with diverse structures and resources, knowing that the data that addressed the question of identity and other emergent issues and intersectionality in relation to learning to teach and teaching would emerge. The meetings occurred bi-monthly, and during the initial meetings, the agenda included a check-in and a reading, graphic, video, or something for them to reflect on and respond to catalyze discussions in relation to the connections between ISE and classroom science teaching and learning.

I also encouraged them to create their own agendas—anything they wanted to share and/or discuss, whether it was a teaching experience, an artifact of their practice, or an activity they were planning and wanted input from the group. For many of the meetings, the check-ins often led to critical conversations around science teaching and learning, whether it was navigating an administrator who was resistant to field trips or discussing parental involvement in the classroom, and it was through these discussions that shared meanings around informal science learning in the classroom in relation to students who have been historically marginalized from STEM emerged. The conversations began with direct enactments of ISE in the classroom but evolved into meanings and practices the Collaborative teachers developed around their notions of what informal science means and affords them and their students.

Agency and Meanings of ISE

I planned the early ILETES meetings around the meanings of ISE and discussed the ways that teachers were using ISE approaches in their classrooms.

The teachers reflected and described things like using science kits, engaging students in hands-on activities, and getting kids outside with "experience and exposure" as important aspects of informal science learning.

One teacher noted, "I see informal science learning as something that's creative and hands-on—it sparks interest and excitement. It uses student engagement and motivation to push them to challenge them to go to a higher level than just understanding the content." Together, we identified the following as characteristics of ISE in the classroom:

- Student choice—The degree to which students can choose what they learn and how they demonstrate learning.
- Centering learning artifacts—The ways that students can demonstrate learning beyond traditional assessments.
- Considerations of place—The contexts of learning: does it only happen outside of the school building?
- Temporality of learning—Having students explore topics over several days or even weeks.

Student interest and motivation were important for the ILETES teachers, and they discussed ISE as a means to leverage these aspects of learning in their lessons and activities. Chen, a high school teacher, wished for:

> …[L]earning to be spontaneous…[for] students to find that the work is interesting and want to learn more…initiate their own learning. It is hard to accomplish in the classroom, but it is what you should do…[create] lifelong learners.

The notion of lifelong learners was an idea that the teachers encountered in their coursework readings and became a motivation in designing science learning. They wanted their students to experience interesting and engaging science both inside and outside of the classroom. Dawn mentioned, "As a teacher, it is my role to motivate them [with] visual learning and hands-on activities, understanding and reflecting, processing. Everyone learns differently." As she described the different learners in her classroom, she noted that it would be important to think about alternative assessments in informal learning. In this, she recognizes that the teaching and learning that she experienced and desired for her students contrasts with the curriculum and assessments that are required that appeared to, at best, narrow science learning affordances for students, if not eliminate it, especially at the middle school level.

Lee was focused on assessments. He was very proud that he had increased the passing rate in Earth science at his school. He worked in a Latinx and Black school in a community with a lower socioeconomic status. His school was considered failing because of the low pass rates on standardized exams. He was both passionate about his subject, Earth science, and even more passionate about teaching. He often spoke about how he saw himself reflected in his students both in terms of their identities and perceptions about them in relation to education. He often brought up that his own teachers "back in the day" did not expect him to amount to anything. He also saw the same sentiment implied towards his students by other teachers in the school. As a white-passing Latino, he was privy to some of the around-the-printer chit-chat that consistently positioned the students as not capable of learning, "I hear these conversations, I just listen because then I know who to trust and not to trust." As I will describe later, trust was very salient for Lee.

The learning culture of his school emphasized standardized assessments, and Lee viewed these as an indication of his effectiveness as a teacher. He wanted his students to pass and excel on the state-mandated exams. As such, Lee discussed changing the environment of his classroom in order to change the way that students interact with each other and with science. His approaches to informal science learning became catalytic in the ways informal science would be defined, adapted, and modified to meet the goals for science teaching and learning of the Collaborative. What also emerged was that their goals were not only teaching the science content in motivating ways but also expanding science learning affordances for their students. They were seeking ways that they could help students build identities and future imaginations where science plays a key role. Through further discussion, it became increasingly clear that the teachers who identified as Black or Latinx with intersections of class, ability, and cultural affiliations saw themselves in their students. Therefore, their teaching also centered on countering the deficit-oriented narratives about Black and Latinx students and redefining the role that science and school play in their lives. With identity being situated and actively constructed and interconnected with the ontological, epistemological, and axiological aspects of being and knowing, I reflected on my own experiences of teaching science while Black and what this meant for me in relation to science and teaching identity. I often shared these experiences with the Collaborative in order to express vulnerability while contributing to discussions about shared experiences.

Lee: Centering Trust

Lee described informal science learning as "any learning that doesn't take place by conventional means...it is not scripted but can have structure...the learning happens by individuals [coming] to understanding by themselves, building self-awareness, but also about group learning where peers help learn." Lee was a catalytic thinker in the ILETES group and designed his classroom so that trust and student agency were centered. He structures his lessons so that students experience agency in learning and rely on peers for support:

> They're in the habit of helping each other; they're in the habit of just taking out their phones. They don't ask me, "What are we doing today?" They look at the weekly tasks and the Do Now, so these habits are developing.

In his classroom, Lee aims for students to develop patterns that center their agency as learners and a classroom culture that values science learning and success. Informal science learning offered him a framework to "[take] himself out of the classroom," with the goal of being more like a "caretaker, where I'm just in the scene, but I'm not really manipulating anything unless I have to unless a student is lost." Lee discussed his stance on learning:

> I think for someone to truly learn, you have to let them develop on their own, and you have to give them tiny little pushes, not a heavy hand. This is why we all disagree with chalk and teach because that's that heavy hand—sit down, pay attention, lecture, lecture. This way, they are learning on their own.

To facilitate this, Lee created a visual system that helps students move through the class activities while helping each other out. Calling this system the "Ice Box," Lee's way of "monitoring everyone without interfering with their interaction." It was a whiteboard with different sections labeled Do Now, Lesson Activity, Exit Slip and Reflection. As students moved through the lesson, they placed their Post-it's with their names in the different sections, so if I'm doing an informal lesson right, I could just be in the back of the room or walking around helping students with the whole lesson (up on the whiteboard) and the students will know what they have to do for that day." The section labeled Ice Box was for places where students were stuck and needed help with moving forward. If a student placed their name there, then either Lee or one of the students' peers would go and help them out. The system allowed him to see who needed help from anywhere in the room and didn't require the students to draw a lot of attention to themselves. His system also helped with students who fall behind, "The student thinks, 'If I'm behind, I'm just going to stay behind'" and does not ask for help "because lots of times, a student may be embarrassed I keep helping them in front of everyone, and [may think] they are the only one behind. Now, another student can come in and say, let me help you, so there's that comfort again coming up in the informal environment." Lee viewed informal learning environments as a space of safety for students who would otherwise continue to fall behind:

> When I come to them, there's a 'heavy hand' of me saying, this is what you are doing wrong; this is how you should do [the task]. Some students may find that a turnoff, and they might stop this process altogether, but if another student comes to them or if they come to that reasoning on their own, then they are going to continue to grow and thrive in this environment. I think that a part of an informal environment is finding those things and wanting to do [them].

Lee also viewed this system as meeting goals of improving his effectiveness as a teacher while appeasing the administration, who may view his methods as being unconventional:

> ...My main goal is not just the strategies. [It's also for] when the administration comes in, and they see something through a different lens—they are going to see all kids on their phones, some engaged in work, some engaged in worksheets, some engaged in activities, they don't know what's going on. So, again, when this is on the board, or on one of my boards, they can see that: they can make the connection that there is something being produced here. This is not just everyone having fun, even though they might be having fun.

Lee's teaching was subversive. From what he described, there seemed to be a strong element of surveillance and mistrust among administrators in his school. He knows that from the window, his classroom would look noisy and chaotic to administrators who, as he described, regularly passed his classroom. He also knew that the apparent chaos in his classroom was effective in that students were on-task and engaged. The Ice Box also created a visual that allowed him to explain to the administrators the structure of the learning that was taking place.

The Ratings

Standardized assessments and corresponding curricula are often in opposition to what teachers desire for their students. In my work with teachers and ISE, they often discuss how the standards truncated their ability to have a range of learning experiences, including field trips and long-term science investigations in the classroom. Katherine McKittrick describes how racism and other forms of oppression underpin academic disciplinary thinking and create "the demand to gather and live with seemingly transparent data…that ostensibly prove that those communities living outside normalcy are verifiably outside normalcy" (2021, p. 4). These standards only narrow educational opportunities for many of the Collaborative teachers' students—Black, Latinx, and urban students. As the dialogues of the Collaborative teachers unfolded, it became evident that their expansivising activities were structured around both meeting the standards and devising ways to expand science learning affordances for their students. As such, the Collaborative teachers discussed the importance of the effectiveness of informal science learning in relation to dictated policies and frameworks (i.e., Danielson, 2007) as well as standards and assessments. These policies outlined what was considered teacher effectiveness and what determined what was assessed. The teachers were both beholden to these structures while actively working to counter or surpass them in ways they believed were best for their students.

According to the New York City United Federation of Teachers (U.F.T., n.d.):

> The Teacher Development and Evaluation System is a multiple-measure system that allows for a holistic assessment of a teacher's contribution to the progress students make. Measures of student learning are combined with measures of teacher practice (observations made by evaluators) to create an overall rating.

Measures of student learning were based on a combination of state assessments and other criteria as decided by a school-based committee of a union representative, teachers, and school administrators. The ILETES teachers often discussed the problems with these ratings—they were highly subjective, and they felt they did not best capture what they did in the classroom.

Beth: Expansivising Against Constraints

Beth teaches in a predominantly Black middle school in a Caribbean community and was expressing frustration with how hard she works, but she receives low ratings, which causes conflict between her and her principal:

> I'm just there to do my job, and as long as my kids succeed, that's all I really care about. But it's like once you [the administration] start to do things that impact the students, it's like we're really not on the same page.

Beth's school emphasized the math and literacy subjects because these are the state-mandated assessments in middle school. Schools were rated according to how well students did on these assessments, and correspondingly, school-based administrators made decisions and policies that prioritized these subjects and the exams. Since science was not assessed, it was not prioritized. Beth was vocal about her feelings that it was a disservice to students, so it put her at odds with her principal.

Since Beth is so passionate about science and believed that exposure to science was necessary for her students, she used her free time to facilitate field trips to different science events around the city. Her informal science enactments went beyond the usual museums, zoos and other cultural institutions to science fairs, open houses, and events at the local college that were meant for undergraduate students. She describes her methods and motivations:

> I took a group of students out last Saturday for free—I don't get paid for that. I took them to Rockefeller University□—every year they have science fairs … so I took the students out [last week], and we had an amazing time. I just want to see my students when they are in those informal learning environments, how they learn, and the opportunities they have. I also see how the things they learned in the classroom come back to them, and I'm just so passionate about that. Seeing them in different environments, they're so excited—they are so beyond excited to just go somewhere different and see different things.

When Beth recounted this experience, she herself was brimming with excitement. With a huge smile and the cadence of her speech increasing, the energy in her voice and body language all indicated that these enactments were important to Beth. "And, these kids love science," Beth emphasized, noting the expectations that Black kids are often positioned as not interested in science, "The kids I took out—they are some of my top students, and they just love science, and they just had an awesome time." The Rockefeller event was developed for students in grades nine through twelve, but Beth offered the opportunity to her eighth-graders. However, she noted she was "a little bit afraid" because she thought it might have been too advanced, but her students demonstrated otherwise:

> ...[T]hey were pretty great. And, in the end, [me] being able to ask them questions [about their experience] and just sitting and watching them on the train and listening to their discussions, you can sort of assess from that time, the majority of them really got it, and they really understood what was going on.

Because these were weekend trips, students self-selected to participate. However, these trips provided opportunities for Beth to observe how students applied the science they learned in her classroom while exposing them to expanded science engagements. These experiences provided enriching opportunities for students and salient professional development for Beth. She was able to watch and listen to them interact with scientists and each other in order to assess their learning and gauge their interests. She understood these kinds of experiences were necessary for getting kids interested in science:

> [In my classroom], I have scientists, and I have thinkers. I believe that a lot of times, kids, like with anything, science isn't fun. Because you just put a book in front of them, they read, and that's it. But when they really go out there, and they feel like scientists, and they really connect with what they are learning and see what they are learning, it's like this discovery—wow.

These trips and her advocacy for science positioned Beth in conflict with her principal since she believed they were taking away valuable time from the assessed subjects. However, the field trips that were allowed and decisions that were made seemed to exclude Beth:

> Last year, I went with the kids because [the administrators] knew that the kids listened to me; I had great [classroom] management skills. [But], now she'll pick chaperones the day of, and then she'll pick teachers who don't even really want to go on the trips, they don't want to really be around...you know, with the kids. And she feels like she's slighting me by not letting me go, [but] you know it's impacting the kids.

In this instance, it is evident that the administrator sees little value in field trips as a valid learning experience but rather views them as entertainment or rewards for good behavior, "The kids just went to see *Age of Ultron* [an Avengers movie] and not to say there is anything wrong with that, but why would you block the kids from actually going to an environment where they can learn [such as] the New York Aquarium, but you have them going to see *Age of Ultron*? I don't know."

The school administrators did not see the same value in field trips as Beth, as both enriching science learning experiences and fun and engaging for students. By deliberately keeping Beth out of the sanctioned field trips, the principal seemed to send a message of power over Beth. Stressed but undeterred, Beth continued to create engaging science learning experiences both in and out of school.

Despite the administrative constraints, Beth and Lee continued to integrate and advance informal learning science in their teaching practice. They believed that these kinds of experiences were beneficial to their students, especially given the existing disparities for Black and Latinx students in schools, as Beth explained:

> In our communities…[we] really need to understand that science is a very important subject and that curiosity is the foundation for many different things. [In] our community, and especially [with] African American and Latinx students, you don't really see them becoming scientists because they're not [often] exposed to it.

Beth found unconventional opportunities for field trips, for example, taking her students to an underground park, "We talked [in the classroom] about mosses…and different ferns and so they have seed plants and seedless plants, so I thought this would be a great way to connect my students not only technology and using science and technology, but an opportunity to really [observe] and draw different plants…connect science and the real world." As this was a new park, for Beth this was also an opportunity for her students to have "a one-up," to experience science in a way that not many others have. In the unique environment of this underground park, Beth engaged her students in complex questions such as, "Why did they choose the plants that they chose?", "What processes make these plants grow?", "What do you notice about these plants [that we talked about in class]?", "What do you notice about the lighting?", and "Why do you think some plants are getting more lighting than others?" The underground park allowed for the expansion of lessons that Beth started in the classroom, only using text-based and printed visual materials. This underground park enabled students to view the integration of science, technology, and everyday lived processes.

In relation to her administrators, Beth's persistence started to pay off, "Last year, I was having a lot of trouble with 'getting' them. But I think now they are so receptive…it's because I really pushed." She believed that they realized her passion came around to support her efforts. They provided her new textbooks, "It was like Christmas time!" and allowed her to go on professional development (PD), "I've asked to go to PDs before, and it was like 'No, you really can't go.' But now I think they are really starting to see that not only that what I'm doing is benefiting the kids, but I think it benefits the school as a whole. The parents came, they were involved, and they asked about me. They are excited, the kids are excited!"

Tara: Expansivising Along a Continuum of Practice

Tara, one of the teachers in the Collaborative, described herself as "teaching along a continuum" of practices from informal to formal. The concept of a continuum allowed the group to view ISE practices as integral to meaningful science learning experiences for their diverse students in the classroom. Tara is an Afro-Caribbean middle school teacher who taught in a predominantly white, affluent district. While most of the students came from the surrounding district, there were students who commuted to the school from adjacent districts that had higher numbers of Black, Latinx, and lower-income students, which also meant they did not have the same academic preparation as their more affluent counterparts. Being situated in this school allowed Tara to observe and experience the deficit notions towards students of color in science and schooling, "It's almost like it's okay if they are struggling." In discussions with other teachers, Tara noticed that the students of color were only brought up for negative behavior and not for academic struggle; she interpreted this as being expected that students of color would fall behind their white counterparts. She also noted higher suspension rates for the few students of color in her school, "We don't have a lot of suspensions, but the students who were suspended last year were mostly students of color." As the only Black teacher in her school, she felt that it was her role to "tap into the students of color," and this meant not only ensuring that they were receiving the same attention as their white counterparts but also designing collaborative activities that ranged from structured to open-ended so that all students had different ways to access and demonstrate learning. This also meant integrating discussions about equity in the context of science topics in the classroom:

> We were talking about energy, and we showed a couple of videos about who has access to what. Even though the video was talking about the conservation of energy and how we can conserve energy, I was showing them examples of my own experiences in Jamaica and how we have experienced full outages. The students were talking about [how this happens in the US].

They discussed the proposal for a dam in the Congo Basin that would generate power for people in Europe, "who has access to what and where?" Tara reiterated. Through integrating her own experiences in a discussion about global access to energy, Tara was able to have her students discuss racial, socio-economic, and geographic disparities while also learning about the laws of conservation of energy. Bringing in the film and personal stories and having problem-based group discussions were aspects of ISE that Tara infused in a "formal" lesson about energy. Along the theme of light and energy, Tara had her students design and build simple solar cookers to demonstrate their understanding of how light behaves. They made tea with their cookers while discussing the use of solar and alternative energy sources locally and abroad. Teaching on a continuum provided a framework for Tara to structure these activities and have these discussions with students so that they increase their awareness of societal inequities while learning the standards-based science topics.

The Physical Space as an Affordance

Both the Urban Advantage and ILETES teachers expanded the role that the physical classroom space played in expansive science teaching and learning. The physical classroom environment affects student learning and morale (Philips, 2014), which is why neglected and unused science spaces, such as in Violet's school (Chapter 5), send a negative message to students, parents, and teachers about the importance of science education. Museums and other informal science institutions (ISIs) recognize the importance of visuality in science engagement and apply many resources and capital to developing and maintaining exhibits that capture the curiosities and imaginations of the visiting audiences. The ILETES and Urban Advantage teachers also recognized the importance of the physical space and purposefully created physical spaces that reflect and encourage ongoing science inquiry. With classrooms being the main site of learning in schools, they should "match your objectives, both in terms of human interaction and your instructional approach" (Philips, 2014). Not only should the seating arrangement be considered, but the organization, visual cues, objects, and resources within the space should also be considered to leverage the affordances of social interactions and the multimodality of everyday learning. For the teachers in both studies, this meant having objects and spaces that reflect learning processes, skills, and science content.

Visiting the teachers' classrooms in both studies, the visual cues—posters, student work, and inquiry prompts along with science-related objects—marked them as science spaces. This visuality provided the needed stimulus to engage students in the mindset of science provided that students know how (or are taught) to access them (Warner & Myers, 2009).

In my own experience of teaching, students noticed when new posters were put up, or new objects entered the classroom. As I did not have my own classroom in the first high school where I taught (I shared with others), many of the objects traveled with me. We reserved the objects, instruments, and/or specimens that we wanted to use for a given lesson. The order was then gathered from carefully labeled storage spots and placed on carts we wheeled to our classrooms. Even having different objects on the cart piqued students' curiosity: "Miss, what are we gonna learn today?"

When I moved to a different school, I set up a classroom dedicated to my science teaching and included the objects I desired. In addition to various posters and student work, I had one classroom pet that captured students, Madagascar hissing cockroaches. They were a parting gift from my previous assistant principal, who moved with me from my Brooklyn classroom to the Bronx, where they captured the attention of even my most disassociated students. They inspired both disgust (most of us have experienced the variety of smaller cockroaches found in our environment as pests) and curiosity (what makes these roaches "pets"? Why do they hiss? What do they eat?) as students encountered them in the classroom. With the same students, we found a planarian in the local urban stream that we then kept in the classroom. Having this brought up many questions and subsequent inquiries, including a neighborhood field trip to explore the health of the Bronx River and observe the various plants—cultivated and wild—that grew in the community around the school.

Creatively Expansivising Affordances

Central to expansivising practices is creativity. Creativity is innately human and compels us to go beyond what is given in order to transform our lives and that of those around us (Stetsenko, 2019). Creative enactments in science education help to frame new ways in and through science teaching and learning. Because of my own interest in dance and the visual arts, I became more interested in the role of creativity in STEM. With creativity being almost always associated with the arts, STEM has become an increasingly popular construct for describing the integration of art and science. My main critique of this approach is that art seems to be used in the service of science, for example, using art to illustrate a scientific concept, rather than as a true interdisciplinary project (where both are used in tandem/intertwined to explore questions and challenges of our times). With creativity leading to new scientific findings, innovations, and programs (Sternberg, 2003), it is important to consider creativity essential to expansivising. Creative teaching and learning approaches, along with opportunities for students to be creative while learning and applying science, should be centralized in the science classroom. Creativity is also a way to center joy and pleasure in science learning (Adams, 2022).

For both the ILETES and Urban Advantage teachers, creativity emerged in different ways. For ILETES, it was how the Collaborative redefined ISE and the affordances they created with that redefinition. In both Urban Advantage and ILETES, creativity emerged in how the science education spaces were organized and used to inspire science engagement. These expansivised spaces allowed students to engage in imaginative, visual and spatial thinking, aesthetic experience, and social and collaborative interactions (Kind & Kind, 2007).

Spatializing Affordances in the Classroom

As a creative and visual thinker, Lee helped to shift the Collaborative both in defining ISE and using the classroom space as an affordance. He described it as "manipulating space", expansivised thinking about the ways that the classroom and other places in the school building could create affordances for creativity in science teaching and learning. Lee shared how he used the table surfaces, windows, floors, and students' bodies to encourage voluntary self-exploration, object engagement, and creative-visual thinking. The examples he provided became affordances for others to think about integrating "manipulating spaces" into their classrooms.

Street Art as Science Affordances

"Middle school students use 'Taggin Up' to visualize what they have learned during the past week. I use this as a formative assessment tool to gauge what the students learn, understand, and can connect to other concepts with this strategy," is how Lee described his practice for a conference presentation. During his preservice education and experiences in several professors' labs, he noted that many scientists used a whiteboard as a space for generating ideas and solving problems. Like other instruments in the lab, it is an essential tool used to share ideas, collaboratively problem-solve, and make the processes visible to multiple participants. To expansivise this practice in his classroom, Lee first used chalkboard paint to transform old, unused lab tables into surfaces for his students to collaborate and co-construct explanations. Using the "tagging", a term associated with the art of graffiti, Lee expanded this practice to other surfaces in his classroom, as he described in a conference proposal:

> The "Taggin Up" method utilizes the classroom's whiteboards, windows, and doors as a way to post up scientific concepts that were learned in the unit and allows the students to share what they learned with the rest of the class. This concept provides a way for room [number] students to have fun and engage them with art and scientific learning.

Students had access to chalk, dry-erase markers, and window markers in order to "tag up" the designated surfaces. For Lee, this process made visible the students' understandings and areas that warrant further instruction and inquiry. This practice also had a trust-building, peer-learning component: the "Buddy Table,"

> The Buddy Table is a multipurpose space. It is a table that is used to assist in lessons and group work by giving the students the ability to construct visualizations of their scientific ideas. The subject of science can be daunting to many; therefore, by using a Buddy Table, the students who are not strong writers but possess strong oral skills can demonstrate their understanding. This table allows my students to explore scientific concepts freely without pressure. The Buddy Table provides rich visual experiences and engages my students in a lesson. The Buddy Table is not only used to assist me in teaching my students; it is used as an opt-out tool as well; my students are given passes to opt-out from other classes. When doing so, they must sit at the Buddy Table to express why they are opting out and working on a science packet.

This Buddy Table serves multiple purposes for Lee, ranging from science learning to socioemotional well-being. Lee described students "opting out" of other classes for a range of reasons, from conflicts with a teacher to being overwhelmed with school and/or home life. Note that opting out was not a free pass to miss a class; rather, students had to articulate why they were at the Buddy Table and were required to do work while there.

Lee's classroom was a haven for many students, especially those with learning disabilities. Identifying on the autism spectrum himself, Lee saw reflections of his own experiences with education in his students; therefore, it was important for him to have a trusting and safe place for students to be themselves while learning science. I recall a video he posted on Facebook with students doing one of his activities—they were all engaged; however, one student stood out for me. He was seamlessly moving between a music keyboard and his table group—playing tunes in between doing his group work. Lee declared, "This is how we do science here!" intuitively understanding that this student needed to have those small music breaks in order to have a better focus on the activity at hand while keeping everyone else in sync.

Learning with the Space and Body

Developing and using models is a practice that is important both in science and in science education as it facilitates deeper understandings of natural and built worlds. Lee led the Collaborative in this respect as well in using the floor and students' bodies to help them grapple with complex scientific concepts. To help enrich learning about contour mapping in Earth science, Lee used masking tape on the floor and wooden blocks. Noticing that his students were struggling with understanding contour and relief maps in two dimensions, he used the space on the floor to help them visualize the transference of a complex landscape onto a flat plane.

As Lee shared his ideas and enactments with the Collaborative, several of the teachers started using their classroom space differently and even extending learning into the school hallways. These enactments further expansivised the ISE/classroom learning culture, turning various spaces and surfaces into active science learning affordances.

Inspired by Lee, Tara brought her students to the cafeteria to play "tug-o-war" and other field day activities in a lesson she entitled "Bodies in Motion." She designed this lesson for a unit that explored how physics is tied to everyday activities. In another unit, students investigated how a ball bearing could complete a series of consecutive uphills and downhills without adding extra energy. In groups, they built roller coasters around the classroom using the floor, tables, and pipes that hung from the ceilings to test their ideas. Tara noted that these experiences, which she described as "informal," allowed her to bring more authentic science learning into the classroom and her students to learn science in a low-stress environment and engage in different modalities of learning.

Similarly, Beth took up the idea of "tagging up" by covering her classroom walls with flip-chart paper and allowing her students to use them to brainstorm ideas, demonstrate understanding, and share knowledge with classmates. She noted that this allowed for increasing student agency, a "student-as-teacher and teacher-as-student" approach to science education.

Collective Identity

During my years of engagement with Urban Advantage and ILETES, it became evident to me the salience of spaces for teachers to build collective identities and develop shared meanings around their science teaching practices. This does not mean that they all end up teaching the same way, but rather it means that they develop the reflexive tools to be able to use, modify and adapt affordances in ways that reflect who they are as teachers and meet the needs of their students as learners. They moved from centering the science content to emphasizing students as learners, as curious, intelligent, and diverse human beings. With an ethos of ISE undergirding their developing agency and identities as teachers, they viewed their students as assets in learning as they were the reason for modifying and creating affordances in the range of ways they did. The ongoing and shared dialogues created the space where their ideas emerged and coalesced, leading to different classroom enactments that were then shared and became points of reflection in the collective. This ongoing and iterative process allowed for the continuous expansion and divergences of affordances, with teachers learning from both successful and not-so-successful enactments. When the Collaborative was asked to reflect on who they were as teachers, they noted that they were creative, problem-solvers, and facilitators more than teachers, "I think we all [do] things that are very hands-on and very project-based," and allow students to create artifacts, beyond tests and assessments, to demonstrate their understandings.

Teachers teaching science in a large, diverse urban district like New York City have the opportunity to draw on multiple schemas and resources, both inside and outside of the classroom, to create the affordances necessary for their students to flourish in science. As such, building relationships and being authentic in valuing and affirming students' identities and cultures is critical in relevant science teaching and learning. Violet described this as being culturally connected to the students:

> Being culturally connected makes the teacher see the student not so much as a student but as a child. The child sees the teacher not so much as a teacher but as another person who wants better for them. And so, it's like one community rather than two communities; it's more like we're doing this together than the teacher wants me to do

it. In that way, we get things done a little faster than we will get it done if we look at it as a teacher/student relationship.

Although Violet's response was to a question regarding her being a Caribbean-American teacher of Caribbean-American students, this sense of connectedness should go beyond having a common ethnic culture. It is about viewing the classroom space as a learning culture where relationship-building and humanizing practices are central. Lee views it in this way:

> Every kid that comes into my class, I have a dialogue with them. I have a relationship with them...you know, I meet them at the door, and I greet them. And at that very instant, whatever, whatever they choose to respond with me, I know we're going to have a smooth day. I know the student is either having an issue with something, and I have to find out. The faster I find out what the issue is, if they're hungry or ... if they didn't sleep or if they have to work...then I have to fix that issue because once I get rid of that distraction and fix that issue, then the rest of my class, rules, management, and instruction become smooth, and that student will buy in. Because they know I care. They know they can talk to me.

Both Lee and Violet acknowledge that the classroom does not begin and end in the school, but rather, it extends into the students' communities and lived experiences. Creating learning environments that center trust, affirming, and relationship building requires resources of collaborative learning, rigorous and creative-science teaching approaches, belief in the positive, asset-based abilities of all students and opportunities for student agency over learning (Adams, 2019; see Chapter 8).

Confronting Neoliberalism in STEM Classrooms

With the Urban Advantage teachers, the social identities of students and themselves did not surface as much in conversations as they did with the ILETES teachers. It is not that they weren't concerned with equity, but rather that the conversations were more focused on the practical aspects of the Urban Advantage initiative—connecting the resources of the ISE to the classroom as well as being responsive to the cultural institutions and the Department of Education mandates. In contrast, the open-ended and longitudinal nature of ILETES allowed for more teacher-centered learning where different dialogues that emerged converged to center racial, ability and socioeconomic equity while still being responsive to the affordances and constraints of their respective schools, their own identities and that of their students. As the researcher in the Collaborative, I became very interested in Nasir et al.'s (2012) notion of racial storylines and the ways that racialized teachers experienced and responded to these for both their students and themselves. Racial storylines describe how prevailing discourses about race get enacted in schools, in particular, the deficit-oriented narratives about Black, Latinx, and lower-income students influencing the learning opportunities that are afforded to them in their classrooms. Black and Latinx teachers are also subjected to these same storylines and experience the deficit-oriented practices two-fold: against their students and against themselves as racialized subjects in schools structured by white supremacy. As such, teachers are made to navigate these storylines both for themselves and their students. They are placed in the position of double resistance as they develop agency in asserting equitable and just STEM teaching and learning for their students.

Furthermore, ongoing neoliberal reforms have enclosed opportunities that Black, Latinx, Indigenous and other students from non-dominant groups have for meaningful science learning. The Urban Advantage and Collaborative teachers often brought up the ways that standardized testing and curricular mandates truncated their efforts and desires for expansivising science teaching and learning. As early-career teachers, the Collaborative reflected on their practicum experiences and described science education as inequitable as they observed in schools of different racial and socioeconomic demographics. Objects and instruments that should have been ubiquitous to science teaching, such as microscopes, were entirely absent in schools in Black, Latinx, and intersectionally lower-income communities. Serena taught students in Tara's affluent school but chose to work at a charter school in the Bronx as she felt that she was "not needed" at the affluent school. However, her experience in the affluent school made her more aware of the inequities when she got her first position at a charter school in the Bronx. In comparison to the largely Black and Latinx district, the students in Tara's school "were on it," meaning that they were at the level expected of middle school students. She reckoned this to be "having more exposure to science early on." In contrast, with her students in the Bronx, right away, she noticed "too many gaps," giving the example of her middle-schoolers not knowing what cells were.

The neoliberal climate dominated by meritocracy and competition has been detrimental to racialized and minoritized students because while they are held to the same "standards" as all, many schools do not have equitable access to the resources necessary for a quality science education. "[B]y misapplying standardized, and efficiency-driven practices favored by businesses to education, neoliberal reforms have constrained the social imaginary of science education within the limits and contradictions of capitalism," excluding minoritized students and constraining spaces of science learning (Strong et al., 2017)."

The ILETES teachers demonstrated ongoing resistance to this enclosure through their expansivising practices. In relating the ways that they enacted these practices back to the question of science teacher identity, I assert that rather than developing a teacher identity per se, they developed repertoires of action for science teaching and learning that resonated with their identities vis-à-vis their students' identities and the desires and science imaginaries that they had for their students. This is an action-identity that is always in relation to and responsive to the various nuances of their teaching contexts. I described the action-identity as Critical Agentic Bricoleur (CAB), meaning "the ongoing augmenting and adapting [of] resources at hand into new science teaching and learning engagements with special attention to attenuating the challenges faced by students [historically] marginalized from science" (Adams, 2019), to highlight the sense of justice and equity guiding their expansivising approaches. These action identities consider learners' cultures, language, and social ways of being in modifying, enacting, and creating affordances. Being and becoming a CAB considers "the sociomaterial entanglements that constitute STEM teaching and learning—the intersections of physical and digital resources and spaces, bodies, languages, and cultures in the conceptual science classroom" (Adams, 2019). However, this is not without the emotional labor, especially for Black, Latinx, and teachers from non-dominant groups, of having to constantly confront, resist, and transform deficit-oriented structures aimed at these same racial groups in relation to STEM and schooling more broadly.

The ILETES research allowed for extending the expansivising framework and making stronger connections between creating affordances and agency. It was also salient to not essentialize teacher identity or agency as these constructs are contextually relational. As described in this and previous chapters, while the teachers all taught in the same system, the schools were diverse in terms of student body, administrative priorities, and geography, amongst other things. Furthermore, each of the teachers brings their own strengths and challenges, as individual humans, to bear as teachers. All of this helps to shape teacher identity and corresponding agency. More importantly, considering equity in science teaching is putting students' and teachers' racial identities in relation to one another. The dialogic, self-selected, and context of the project—that it was neither in association with a museum nor school—allowed for more frank discussions about race and other social factors that influence what happens in the classroom. With the focus on identity, teachers were also inclined to share their experiences with their own social identities vis-à-vis science teaching and learning. The teachers accessed, adapted, and transformed science affordances to advance equitable science teaching and learning in response to the racial and social identities of both their students and themselves.

Expansivising Social Justice

Teachers encountering new schemas and resources engage in a Use-Modify-Create (UMC) process in order to integrate new affordances in the classroom. UMC is a nonlinear process of bricolage; as teachers learn about and use new-to-them resources and schema, they integrate them into their existing repertoires of practice, also changing these repertoires in the process. Agency evolves as teachers become more reflexive, fluid, and responsive in the ongoing process of adapting and creating new affordances. The affordances become acts of social justice when teachers' awareness of racial and social inequity—both from their personal experiences, understanding of the historical contexts of schooling and observations of inequity across different schools—and dismantling these inequities becomes central to who there are as teaches (identity) and their corresponding practices (agency).

Expansivising teacher learning moves beyond the introduction of teachers to ISI resources to meet the needs and identities of students with an eye toward expanded future imaginations, especially for racialized students. Justice-oriented expansivising emphasizes students' identities their intersectional social identities, including their positions as young people with a strong stake in experiencing futures that are socially just and center planetary wellbeing. For students to develop scientific-creative-agency, they require learning spaces where their humanity is celebrated. Educators need to reflexively ask:

1. In what ways is science learning important for my students?
2. What futures could they imagine for themselves in/with science?
3. How can teachers use the affordances available to help shape these imaginations?

Teaching on a continuum is being intentional about leveraging the affordances of formal and informal science in order to enact student-centered, identity-responsive, and equity-oriented science teaching and learning. Good science teaching also requires considering a constellation of practices that use, modify, and create material and conceptual affordances toward expansive science teaching. Good science teachers create bricolages of resources and schema in order to create and expand affordances for science learning. They become critically aware of inequities through reflecting on their own experiences as racialized subjects, their experiences in different classrooms and challenges with existing structures that truncate meaningful science learning. They actively resist with their bricolages and expansivising practices. In this way, Lee leverages the informal interactions of his students in building trust and communication, and Beth accesses the affordances of students' out-of-classroom interests to make structured connections to science content and processes. Tara uses her ISE approaches to highlight her racialized students while engaging all students in meaningful, relevant, and justice-centered science education. Through these enactments, the Collaborative teachers were bricoleurs—adopting expansivising stances toward science teaching and learning.

Dialogic Spaces for Expansivising Practices

> Jah come to break downpression, Rule equality,
> Wipe away transgression, Set the captives free.
>
> -Robert Nesta Marley

I start this section with prose from Bob Marley because, to me, it speaks to the captivity of unexamined science education and unchallenging the "transgressions" of being Black vis-à-vis science and schooling. However, bell hooks' notion of transgression calls us to go beyond what is possible and view the classroom as a radical place of possibility. In such radical spaces, transgressions of identity become assets and resources for science learning. Centering students' identities—who they are and the questions they ask about the world—is integral to affording expansivising science learning and teaching. The teachers described in this book had expanded visions of what is possible in science education, and their notions of ISE allowed them to begin to realize these possibilities in their classrooms. In the expansivising place at the nexus of ISE and schools, teachers transgressed expected science teaching and learning norms for many urban students. They transformed the transgression of being Black and Latinx in STEM by using, modifying, and creating affordances that allowed their students to engage in science learning in meaningful and affirming ways.

In science teaching and learning, we often center on knowing the content as the most salient. While it is important to have a deep knowledge of and curiosity about science and the ability to meaningfully engage students in relevant content and topics, being a transgressive, expansivising science educator requires much more than this. It requires educators to be critically reflective about STEM in relation to society; this means that STEM educators, across educational levels and settings, have the foundational awareness of the ideologies that shape science vis-à-vis society, how this manifests in STEM education, and how we can best educate learners to transgress the oppressions that are explicit and make the invisible oppressions visible in order to eliminate those.

Collaborative, participant-centered, research-to-practice projects are critical in transforming STEM teacher education. For these kinds of projects to be the most effective in teacher learning and agency, they should be longitudinal—offering teachers the time and space to engage in ongoing cycles of reflection, dialogues, design, and enactment towards the development of repertoires of practice—actionable practices they can draw on both in the planning of lessons and activities and the moment during unfolding enactments in the classroom. We first described this as Spielraum: the development of practices that are anticipatory, timely and appropriate to situations that arise in the classroom (Roth, Lawless & Masciotra, 2001). We described how reflecting on teaching experiences to develop actions for change was important to developing Spielraum (Adams & Gupta, 2017). Repertoires of practice extend this by highlighting not only the physical and pedagogical moves that a teacher makes in teaching but also the ways that identity shapes how one approaches these moves. In other words, the same or similar practices could be driven by different visions and, therefore, both enacted and experienced differently by the teacher and learners. This does not place a value judgment on the enactment; it just establishes the contextual nature of repertoires of practice.

Expansivising spaces provide opportunities for dialogues about practices to be shared and new practices generated as participants share and make meaning of how different resources can be adapted and used to teach science. Although the dialogues were unstructured, they fell under the theme of using informal science resources and approaches to improve urban science teaching and learning. This allowed for different topics and issues at hand to emerge and shape the dialogues and actions teachers developed to address challenges and inequities.

Teachers were encouraged to share their successes and frustrations with developing and implementing ISE-related practices in the classroom. This enabled teachers to empathize with enactment issues while also viewing constraints as opportunities to expand their ISE practices as a group. The open-ended, question-oriented, self-directed approaches the teachers valued in their ISE learning experiences were also mirrored in these unstructured dialogues. The informal science teaching and learning theme prompted a variety of discussions and created more space for teacher knowledge sharing and co-generation.

Notes

1 A national initiative to recruit, prepare, and retain "highly effective" mathematics and science teachers in designated high-need school districts. The New York City Department of Education schools fall under this designation, as there are often teacher shortages in key subject areas, especially in STEM, and at least 30 percent of the students, city-wide, have family incomes that fall below the poverty line.

References

Adams, J. D., DeFelice, A., & McCullough, S. (2022). Teacher-learning, meaning-making, and integrating ISE practices in diverse urban classrooms. *Connected Science Learning, 4*(4), 12318651.

Adams, J. D. (2019). WhatsApp with Science? Emergent CrossActionSpaces for communication and collaboration practices in an urban science classroom. In *Emergent practices and material conditions in learning and teaching with technologies* (pp. 107–125). Cham: Springer.

Danielson, C. (2007). *Enhancing professional practice: A framework for teaching.* ASCD.

Kohli, R. (2014). Unpacking internalized racism: Teachers of color striving for racially just classrooms. *Race Ethnicity and Education, 17*(3), 367–387.

Kind, P. M., & Kind, V. (2007). Creativity in science education: Perspectives and challenges for developing school science. *Studies in Science Education, 43*(1), 1–37.

Luke, A. (1995). Text and discourse in education: An introduction to critical discourse analysis. *Review of Research in Education, 21*, 3–48.

Martin, S. (2006). Where practice and theory intersect in the chemistry classroom: Using cogenerative dialogue to identify the critical point in science education. *Cultural Studies of Science Education, 1*(4), 693–720.

McKittrick, K. (2021). *Dear science and other stories*. Duke University Press.

Mocker, D. W., & Spear, G. E. (1982). Lifelong learning: Formal, nonformal, informal, and self-directed. Information Series No. 241.

Phillips, M. (2014, May 20). A place for learning: The physical environment of classrooms. *Edutopia*. Retrieved from https://www.edutopia.org/blog/the-physical-environment-of-classrooms-mark-phillips

Roth, W. M., Lawless, D. V., & Masciotra, D. (2001). Spielraum and teaching. *Curriculum Inquiry, 31*, 183–207.

Tobin, K. (2006). Learning to teach through coteaching and cogenerative dialogue. *Teaching Education, 17*(2), 133–142.

United Federation of Teachers. (n.d.). Your guide to the teacher development and evaluation system. Retrieved from https://www.uft.org/sites/default/files/attachments/evaluation-guide.pdf

Park, E. L., & Choi, B. K. (2014). Transformation of classroom spaces: Traditional versus active learning classroom in colleges. *Higher Education, 68*(5), 749–771.

Phillips, M. (2014, May 20). A place for learning: The physical environment of classrooms. *Edutopia*. Retrieved from https://www.edutopia.org/blog/the-physical-environment-of-classrooms-mark-phillips

Stetsenko, A. (2019). Creativity as dissent and resistance: Transformative approach premised on social justice agenda. In *The Palgrave handbook of social creativity research* (pp. 431–445). Cham: Palgrave Macmillan.

Warner, S. A., & Myers, K. L. (2009). The creative classroom: The role of space and place toward facilitating creativity. *Technology Teacher, 69*(4), 28–34.

· 7 ·
CHALLENGING "SCIENTIFIC" RESEARCH PARADIGMS THROUGH SOCIOCULTURAL LENSES AND DIALOGIC METHODOLOGIES

Teaching, learning to teach, and learning are relational and dialogic processes. Through continuous interactions between people and contexts, people learn and subsequently shape their individual and collective identities. In science teaching and learning, this is further complicated by peoples' relationship to science, their perceptions about science, notions about who can learn and "do" science, and the science curricular content itself. These are all implicated in the ongoing dialogic processes in science teaching and learning contexts. As qualitative and critical-oriented researchers, it is germane for us to employ research methodologies that mirror the teaching and learning processes that happen in a given context. This has the potential to elucidate social life's nuances in ways that solely conventional methods, such as interviews and surveys, may be missed. In this chapter, I will describe my journey through the dialogical and participatory methodologies and corresponding analyses and discuss how this enabled me to expand my ideas about teacher identity in relation to practice and the ways that I approach teaching and learning research.

Researching Teacher Identity

In the study I describe in this chapter, I wanted to learn about the relationship between new teachers' identities and the contexts in which they learned to teach. Informal Learning Environments and Teacher Learning for STEM

(ILETES) focused on teacher learning and Informal Science Education (ISE). I recruited new teachers in their first three years of teaching who participated in credit-bearing courses that took place either in informal settings or integrated informal science resources into their learning experiences. In the United States, ISE largely refers to science learning in science-rich cultural institutions such as museums, science centers, zoos, and botanical gardens. More broadly, it refers to science learning beyond the school's temporal and/or spatial boundaries, including after-school programs, place-based settings, and cultural science learning (NRC, 2009; Adams, 2006).

I began the research with initial conceptions about the relationship between teacher identity and learning to teach in informal settings. In prior research, I learned that the affordances of informal environments, such as having access to multiple and diverse audiences, interactions with authentic objects, and opportunities for self-reflection, contributed to student-centered and equity-focused notions of teaching (Gupta & Adams, 2010). This research was extended to examine a preservice teacher clinical experience in a museum setting where we focused on learning more about ways to support developing teacher identity. Our findings pointed to the importance of recognizing teachers as sociocultural beings who bring beliefs about students, learning, and their role as teachers into the learning-to-teach environment and the salience of providing experiences that allow for interactions with diverse learners and opportunities for reflection on those interactions in developing emerging identities that focus on equity and engagement (Adams & Gupta, 2017).

As these studies focused on preservice teachers, we did not have the opportunity to understand their enactment of teaching once they entered the classroom. However, these assertions contributed to the present study, in which I endeavored to learn how these informal science learning experiences transferred into the classroom. My initial overarching research question mirrored this reality, "In what ways does teacher informal science learning influence teacher identity?"

Definitions of Learning Contexts

With this in mind, I first revisited definitions of informal, formal, and nonformal science learning to establish a baseline of the meanings of these terms according to extant literature. With informal learning often tied to times and places outside of the classroom, I noticed that teachers in my research tended to position informal science in contrast to classroom science, using terms like "open-ended," "fun," and "creative" to describe the former and "rigid" and "structured" to describe the latter. Informal and formal learning were often dichotomized, while the reality is that it is hard to fit any learning context into one category as there are overlaps—there are elements of formal, informal, and nonformal learning environments.

Understanding the aspects and interrelationships of learning approaches as situated in the contexts in which they occur helps guide cross-contextual teacher learning (Colley, Hodkinson, & Malcolm, 2002). For this study, I built descriptions of learning across the continuum based on Mocker and Spear's (1982) definitions of formal, informal, and nonformal learning. This allowed for the beginning of articulating elements of learning and how they were presented in the teachers' discussions, as well as providing a common language for the teachers to describe their practices concerning learner agency. For me, it was important to find that connection between dialogue and practice and begin to think about the different contexts of learning in terms of a continuum encompassing a range of practices.

At the time of the study, I was a collaborator at a human ecodynamics field school in Barbuda, a small, low-lying island in the Eastern Caribbean. Since the local economy was agriculture-based and food security was a social issue, much of the field school centered on different, transdisciplinary learning activities centered on farming and sustainability. The field school was a formal, credit-bearing course; however, the teachers were able to learn through different activities along the continuum of formal to nonformal learning. For example, as they were learning how to transfer plants from a nursery into the ground, they had dialogues with different participants that ranged from learning about the uses of the plants from elders to how Barbudan teachers used the natural environment to teach science. These experiences provided different ways for the teachers to learn science and how to teach. The Barbuda field school provided them with experiences to later reflect on and think about how to integrate both the knowledge and approaches in the classroom. One of the teachers built an aquaponics tank in her classroom, like the one she interacted with in Barbuda. She was able to use this setup to discuss the curricular science content while sharing her experiences in Barbuda and the realities of global food security issues. This also allowed her students to have various ongoing learning experiences with the fish, plants, and technology incorporated in the aquaponics tank.

Sociocultural View of Identity

Sociocultural frameworks are germane for elucidating how learners engage in complex socio-material contexts, including interactions with different people, ideas, objects, and seen and unseen structures. Thus, as I examined identity, I looked for frameworks that explained how contexts shape identities and what happens to these identities when people move through and interact in different spaces as learners and teachers.

I began by defining teacher identity, which is how a teacher represents herself through practices and positioning (Avraamidou, 2014). That is, what a teacher does and how she positions herself in relation to others in given contexts (colleagues, students, administrators, parents, etc.) will reveal a particular way of being—an identity. Connecting identity and agency was also salient in understanding the ongoing process of learning to teach: "agency and identity develop interdependently and as an ongoing process of learning not only who is the self who teaches, but also who is the self in relation to others who teach, learn and learn to teach science" as well as the self in relation to the contexts in which one teaches (Adams & Gupta, 2017, p. 4). In this sense, agency is the belief in self-capability in the social project of teaching. As identity is an ongoing process, agency entails the ongoing project of learning to be a teacher by learning practices and ways of being a teacher. Learning to teach is learning how to use, adapt and integrate a range of conceptual and physical resources in teaching. Furthermore, a creatively-agentic teacher is able to integrate existing resources (or lack thereof) in novel ways. This is an agency in teaching, and as a teacher experiences success (and is able to learn from failure), they experience more agency in a particular orientation towards teaching—including attitudes about student learning. This shapes who they are in the classroom and the resources they seek out to allow this identity to progress.

Although this notion of identity is largely described as an individual project, it is important to realize that it never is. Identity is always the self in relation to others or the self/other dialectic—in teaching and learning, I would also add *place* to that equation—in that all three aspects are inter-relational and cannot exist or evolve without the influence of, or influence, the other.

Contexts of Learning to Teach

The teachers in the study all participated in learning experiences (credit-bearing) that integrated an informal science learning approach. These courses were designed to familiarize teachers with pedagogies and practices of informal science learning while introducing them to different places and spaces to teach and learn science. The courses tended to emphasize an inquiry-based and/or place-based approach. In general, there were two designs:

City-as-Classroom Approach: As explicated in Adams, Miele, and Powell (2016), this approach centers on place-based experiences in science learning. Although these courses focus on learning science content both for earth and environmental science majors and preservice teachers, the pedagogical approach emphasizes experiential and project-based learning. For example, in a course on hydrology, students learn the topic through visiting a museum exhibit about hydrology, field trips that explore places that are central to supplying water to the city and water treatment plants, in addition to traditional classroom-based activities about the water cycle and other hydrology-related topics. Thus, this course *models* how to integrate out-of-classroom resources in science learning.

Museum-based Courses: These courses take place in a science-rich museum and are designed around using the museum's resources to learn and teach science. Teachers are immersed in learning in the museum's halls and exhibits while learning the pedagogical approaches of learning in the museum, for example, learning how to use questions to engage students in observing and learning with dioramas and object-based learning (Adams, 2006).

In ILETES, it was important to learn how teachers took up these informal science resources: how they define informal science learning and how they interpreted their approaches in their classroom to meet their teaching and learning goals. These aspects are salient in shaping and maintaining a particular stance to teaching—a teaching identity.

Centering Dialogues

I used dialogic approaches as they allowed me to learn about the teachers and their developing/expanding identities, and these dialogues allowed them to learn from one another while building collective meanings and practices around ISE. The dialogues happen when they talk about their teaching and how they speak about the artifacts that they share with their peers. These artifacts could be anything from pictures of things they have done in the classroom, descriptions and pictures of field trips they have taken with students, worksheets and student work, and even workshops they hear about to share with the rest of the group. We (researchers) documented how they talked about their teaching, learning, and students in conjunction with the artifacts they brought. The meetings were all audiotaped. We first videotaped but found that the situation of the room made it challenging to capture everyone, and for this study, the dialogues were more important than the physical gestures. My iPhone came in handy for this because I always have it, and the audio is of good enough quality for both transcriptions and revisiting the dialogues.

The bi-monthly meetings lasted anywhere from one and a half to two hours. During the first year, they were open-ended to allow themes to emerge in the teachers' discussions. During the second year, we called the space "Collaborative Teacher Inquiry around Informal Science Learning and Science Teaching in Urban Classrooms" because the teachers valued the collaborative aspects of the meetings and how they contributed to their professional development. As researchers, we used these dialogues to interrogate and expand our initial concept of identity and describe different places and spaces that teachers use in their teaching concerning the framework and informal science learning.

Engaging with the Data

Constructivist Grounded Theory

While I started with a constructivist grounded theory approach (Charmaz, 2005), I moved towards a more iterative approach that became a dialogue between the theory, the data, the context, and the participants. I knew that my positionality would influence the research process and my interpretations, so I felt that dialoging with the data was important to ensure that different perspectives were accounted for and that the emergent findings were useful to the research participants and contexts.

My history and experience as an educator, experiences in the same contexts as my research participants and valuing collaborative approaches all influence how I approach research. While I cannot remove nor ignore my own lens, what I focus on in the research, collaboration allows me to learn from others' perspectives and expand what I pay attention to when I interact with the data. Working with a research team, considering research participants as collaborators, and sharing findings as they unfold with different groups of participants offer multiple perspectives on the data and feedback on unfolding findings. This also allows for relevant participants to benefit from and access the research at different phases, not just in the publications (which benefits a very narrow range of participants).

In terms of data analysis, I have been moving more towards intuitive reasoning, a nonsequential and iterative way of processing information that accounts for cognitive and affective aspects of direct knowing (Sinclair & Ashkanasay, 2005). An intuitive process acknowledges that one's position or location in relation to the research context is a salient way of knowing and learning more about the research theme or question at hand. As a reflexive and iterative process, it requires multiple passes through the data, with each pass layering the emergent themes, contradictions, and questions. The nonlinearity of intuitive analysis is aligned with everyday decision-making and allows for the emergence and mergence/coalescing of data and theory in ways that afford the framing of participant-centric learning and social life theories—theories that are more resonant with how social life is experienced.

Intuitive and dialogic approaches to research and data analysis are best done in the context of a research collaboration with each participant, bringing their positionalities and experiences to make sense of the data. For example, I shared my unfolding findings with the ILETES teachers in order to get their take on the data, which I will later describe. I also shared my perspective on the data in my regular research meeting, during which my collaborators shared their emergent findings. These meetings allowed for important and nuanced discussions around the meanings and relevance of different theories, operationalization of terms, and implications for practice to be generated.

For the work I describe in this chapter, I considered the narratives of the teachers' stories they tell about themselves in the classroom. Stories we tell about ourselves reveal who we are in relation to others, contexts, and practices. Teachers engage stories about themselves as teachers—these dialogues often include what they do and why they do what they do. I found the notion of counterstory in Critical Race Theory useful for analyzing and describing teacher identities in this work as it centers the experiential and embodied knowledge of the teachers (Martinez, 2014), who were mostly people of color and allowed for an elucidation of how they used their notions of ISE to create equitable and just learning opportunities for their students, who were mostly students of color. Counterstorying allows for the centering and retelling of the teachers' stories to demonstrate their agency in the classroom despite administrative and structural challenges that negatively affect science education for Black and Brown urban students. Counterstories can serve to build community, collectively challenge norms and assumptions, deepen understanding of structures underlying inequities and expand vision of possibilities beyond given realities (Solózano & Yosso, 2002).

The stories teachers shared during the cogenerative dialogue meetings helped me gain a deeper understanding of their identities vis-à-vis their students and the contexts where they learned to teach and teach. I used the stories to develop "identity vignettes" to show the relationship between their understandings of ISE, their unfolding teaching practices, and their teacher identities. This was not meant to essentialize nor define types of teachers but rather show a more dynamic relationship between identities, interactions, and contexts of teaching and learning. I presented the identity vignettes to the participants to see if they made sense in relation to how they saw themselves.

Lee, a middle-school Earth science teacher, referred to the vignettes as "archetypes." Immediately, the tarot major arcana came to mind, so I looked up "tarot archetypes" and found, "Technically speaking, an archetype is a primal pattern of thought..." popularized by Carl Jung; "... we're all pre-programmed to look for archetypes in our everyday lives because they serve as a framework for our understanding of the world" (Kenner, n.d.). I thought this description apropos for making sense of what I was gleaning from the data and my interactions with the teachers regarding their identities. For me as a researcher, these archetypes provided frameworks for understanding the ways that teachers make meaning of their relationship to informal science practices in relation to who they are and how they view themselves vis-à-vis their students.

Emergent Understandings of Teacher Identity Development

To understand teachers' conceptions of ISE, I first looked across data sources to understand how they, as a collective, described informal science learning. First, they described it in terms of purpose and the pedagogical role informal science learning played in their practice. They viewed it as a way to motivate, energize, and engage students, encouraging spontaneity and the love of learning. They also described different spaces and activities in the classroom as "informal," and these tended to be places or activities that were collaborative, student-centered, and multimodal. As a part of defining and articulating what ISE or learning meant for the teachers, I introduced to them the strands of learning outlined in the National Research Council's *Surrounded by Science* (2010). This publication was meant to provide a basis for starting a conversation in the field about the meaning and modes of informal science learning for practitioners. When the teachers reviewed these learning strands, they related them to classroom-relevant words and concluded, "This is just good teaching." In other words, the informal science learning strands or practices were not much different from their understanding of what good teaching looks like. Although they were very specific in defining and categorizing what is considered informal and what is not, the reality is that there is a blurred boundary between the two, a theme that became more evident to teachers as the research unfolded. Furthermore, although there were common definitions, how ISE was enacted in classrooms demonstrated that the practice took on different meanings for the different teachers, largely depending on who they were as teachers and the goals they had for their students, both as science learners and as social beings.

The Archetypes

The Striver

Wei's own educational experiences centered on memorization and recitation. He did not have the same level of immersive ISE experiences as the other ILETES teachers but rather encountered it in an online course offered by the American Museum of Natural History. These courses were designed to deepen teachers' content knowledge while structuring inquiry-based activities and project-based learning as a part of the course approach. Because he did not experience the degree of modeling and practice the other teachers experienced, his understanding of ISE was emergent and developed as he participated in group dialogues and shared activities. He viewed informal science learning as the opposite of his own learning experiences and aimed for "learning to be spontaneous…[for] students to find that the work is interesting and want to learn more… initiate their own learning; it is hard to accomplish in the classroom, but it is what you should do… [creating] lifelong learners." As such, he inserted moments of hands-on activities and challenges in the more lecture-oriented environment; for example, in a physics unit, he provided students with an assortment of materials with the challenge of building a bridge that would support the most amount of pennies in a cup. For Wei, informal science learning was a hands-on activity and student choice. He noted that he defaulted back to lecturing at times because it was his comfort zone, and he felt it was the best way to get information across to students. However, he described another small, inquiry-based activity he tried out with students each time we met. It seemed that the ILETES group offered him the space to share, reflect, and receive peer feedback and ideas and move away from his lecturing comfort zone. He hoped to one day have a practice centered on student choice and problem-based learning.

The Orchestrator

Lee came into the meeting one day with a clear plastic file box. Within this box, he had folders, each labeled with a student's name. When he opened one of the folders, he displayed a page that looked like a bingo card—in each box was a carefully planned activity from which students could choose. He described that he planned a unit around these activities, with each of the activities allowing students to "touch and manipulate things" to learn Earth science concepts. Within this structured space, his students engaged in science in very personal and individual ways.

Lee self-describes as high-functioning autistic and had an Individualized Education Plan (I.E.P.) in primary and secondary school that allowed him to get special education services throughout his schooling. For Lee, informal science is about having students engage in science at their own pace and in their own ways. However, he did not focus only on individual activities; he structured his learning environment to foster collaborative learning and peer support. He devised very structured ways of designing learning experiences for his students that only he could accurately describe, but I will try to reproduce them here. For example, he had a blackboard with different squares, one labeled "The ICEBOX". He described the icebox as a space for student support—if a student was working on something that he or she did not understand, they would put their name in this box, and either Lee or one of their peers would help them. Once they understood, they removed their name from the icebox and moved on to another activity.

In Lee's practice, I realized two contradictions. The first is the contrast of a rigid structure and free-flowing movement and choice. He had a specific way of organizing the board that guided students' approaches to the activities; however, they had choices in sequence, pace, and opportunities to ask for help. This structure afforded the engagement of his learners with a range of abilities and ways of learning.

Lee had a very creative and seemingly open-ended approach to how he structured his learning experiences; however, he is also very beholden to the statewide assessments, known as Regents Exams. Schools often assess teachers by how well their students fare on this exam; for Lee, this is how he self-assesses the quality of his teaching. Thus, his identity as an Earth Science teacher is also structured by this exam and how students performed on this exam. However, Lee used this constraint as another opportunity to expand the spaces for student learning. For example, he employed the street art process of "tagging" for Regents Exam review. Students use dry-erase markers to draw science concepts on the windows, and then Lee facilitates a gallery walk through the "art" for review. He also had desks with blackboard-painted surfaces that allowed his students to "doodle" concepts as they grappled with science learning—the spaces in his classroom, in essence, became a canvas for student learning.

For Lee, the "informal" was about creating multiple learning spaces and opportunities for his students. As an orchestrator, Lee creates, monitors, adjusts, and assesses opportunities for student learning in ways that are student-centered and resonate with his own learner identity and desire to be a successful Earth science teacher. Lee was also a transformative-thinking force within the teacher research group as several collaborative members incorporated his ideas into their classrooms.

The Advocate

Feeling constrained in a school that emphasized math and English Language Arts (ELA), Beth felt that she was in a constant battle with her administration for science. She is passionate about science and believes that science is a way of expanding her students' knowledge about the world and futures that they imagine for themselves. For her, informal science was connected to out-of-classroom learning. However, for Beth, this goes beyond the traditional trips to the museum and extends to the different places and opportunities that the city has for students to engage in science.

Because of her strong passion and belief in exposure to science, Beth often used her own free time for field trips. Being in an assessment-focused school, the administration was not generous with allowing field trips because they took time away from test preparation. This constraint caused Beth to find opportunities for science learning and exposure in spaces often not structured for school trips, such as community Science Saturdays at a premier research institution and conferences at local colleges. At these events, students were able to interact with scientists, faculty, and postsecondary students, thus affording them perhaps a more authentic and engaging experience than solely visiting museums. Furthermore, Beth noted that these trips served as "informal" professional development for her—watching her students interact and listening to the questions that they asked—allowed her to observe and understand student learning in different settings.

I had the opportunity to visit with Beth when she took her students to an environmental science conference at a community college. I was first impressed with the number of students who came on a Saturday; she also had a new teacher from her school with her, as well as some of her students. I was also very impressed by the quantity and quality of questions that her students asked; they were on par with the college students present. She mentioned that she prepped her students beforehand and gave them extra credit for asking questions, which, I thought, led to very thoughtful and complex questions.

As a teacher with a similar racialized background to her students, she believed that it is important to expose her students to science in as many places and ways as possible. She understood the stereotypes around Black people and science in the United States and worked to resist those stereotypes both for herself and her students and give them opportunities to create alternative identities. She addressed the students as "Dr." in her classroom and on her field trips, allowing them to "try on" a science-related identity. Also important to her identity and practice was being connected to the ILETES group. Whenever she took her students to a new place, she often texted me pictures of their journeys. This is part of identity confirmation—having me acknowledge what she is doing; Lee included me in his social media review sessions with his students. These were ways of confirming their identities, through recognition by me, their professor, as teachers who value and create informal science learning opportunities for their students.

The Manager

Luis was in a school on an island in the harbor that focused on marine and harbor-related careers. It is different from the schools of the other teachers in the group in that the emphasis is less on standardized tests (although students still are required to take them) and more focused on marine-related careers. For Luis, informal science was a way of getting the vocational-technical experience while learning science, for example, learning how to navigate the harbor while learning about tides, estuaries, and underwater geography. He mentioned that most of the higher-level harbor jobs (boat captains, pilots, tugboat captains, etc.) went to people from outside the city and were mostly white. His goal is to have more diverse and local New York City students adequately prepared to take on these jobs. Luis focused on learning, which allowed his students to gain work skills and experiences for those jobs while learning science content and practices. For example, as a Navy veteran, he knows knotting is a critical boating skill and important to marine safety. He used knotting to teach about friction and loads in physics.

Luis also viewed informal learning as something unexpected that happened in his classroom, citing an experience where the school acquired a squid specimen that they were immediately able to use in his classroom. Similar to Beth, participating in the ILETES group was important in confirming his identity as an "informal science" teacher. He consistently emailed the group about ISE professional development opportunities. When historic vessels visited the NY harbor, he actively created and shared lesson plans, ideas for subject-area integration, and information for school trips to visit the vessels.

Meanings of ISE

ISE takes on different meanings for different teachers. While they associated common practices with informal science learning, for example, inquiry-based learning, hands-on activities, and field trips, they diverged in the essence of meanings; in other words, what it meant for them to enact informal science learning in their classrooms. The teachers enacted ISE in ways they saw best met their students' needs as learners and resonated with their identities as educators. For example, Lee, who identified as a neurodiverse learner, focuses on developing a range of activities so that his students all have equal opportunities to learn. Luis, as a naval veteran, views informal science learning as a way for students to obtain career-related skills while learning science.

Collective Meanings/Individual Identities

The dialogic approach provided a space of growth and support for the teachers as they navigated their first years of teaching and developed their teaching identities. They often spoke about how important it was for them to come to the meetings, both in terms of debriefing and reflecting on their own teaching experiences and learning from others' experiences. Through this process, the group developed collective meanings around informal science learning while enacting practices in relation to their individual identities and the diverse needs of their respective classrooms. They also gained an understanding of how their respective schools influenced their teaching—from the resources available in their building to the ways that administrative mandates either afforded or challenged their teaching identities and goals. For instance, being situated in a school focused on math and literacy forces Beth's informal science teaching and learning opportunities to spill over into her time. Because of her belief that exposure to diverse science learning experiences was important for her students, she was willing to sacrifice the time. Because Lee's school emphasized Regents test exams, he focused his informal science approaches on the exam content and measured his successes according to his students' passing rates.

In this study, I played the role of both the researcher and the facilitator. In her work with teachers of color, Rita Kohli (2014) refers to "reciprocal vulnerability," where the researcher shares her experiences with the phenomena to create a space of trust and collaboration. As someone who taught in the same public school system as a science teacher, I was able to share my own experiences—challenges and successes—with teaching and navigating administrative structures and logistical challenges around enacting transformative teaching in the classroom. As a teacher of color, I was able to share my own experiences with race in relation to my role as a teacher, including my own perceptions and commitments to teaching students of color. Furthermore, as an educator who values informal science learning and his experiences as a teacher educator in a museum, I was able to both share best practices and, through my collaborations, offer entry for them and their students to different places and experiences. While in some instances, it may have produced biases on my part as a researcher, having shared experiences with the ILETES teachers allowed us to build a collective where teachers felt free to share their experiences. As a new teacher, I also played a role in their induction by creating an informal space for them to gather and talk about their teaching with me; as an experienced educator, I was present to offer some insights on their reflections.

Bricoleur/Age as Teaching Identity and Approach

The term "polyphonic bricolage" describes the continuous process of creating new culture as people interact with new resources and environments (Schmidt, 2008). Teachers also engage in this process as they encounter new places, spaces, and resources for teaching. Through these encounters, they create "new" ways of enacting science teaching and learning in their classroom, thus transforming and expanding the learning opportunities available to their students.

It is important for teacher educators to keep this notion of bricoleur in mind when teaching teachers. For example, informal science teacher learning experiences tend to focus on using resources, such as exhibits and objects and/or approaches, such as inquiry, object-based learning, or project-based learning. While exposing teachers to resources and approaches is important, it is also critical to facilitate discussions and activities about using and adapting to transform learning environments. This is best done through teacher dialogues—novice teachers having dialogues with experienced teachers about how they adapt and transform given resources in the classroom to meet teaching and learning goals. In this way, resources become less rigid and more expansive in how they can be shaped to fit the learners and learning needs at hand.

We also need to consider what structures teachers need to engage in this kind of teaching. In other words, what supports are needed in order to give teachers the space and practice to do creative teaching enactments with the resources they acquire? In this research, the long-term nature and collaborative nature of the group provided the space for teachers to not only learn from each other but also define for themselves and each other the meanings and affordances of informal science learning. They created collective meanings and individual practices to meet the needs of their different teaching contexts. I struggle with how to maintain such a group outside of the grant (as the funding provided food and gathering space); however, I believe that these spaces could be created within schools or, more specifically to this research, informal science spaces; perhaps monthly drop-ins where teachers could come and get judgment-free feedback on their teaching and suggestions for activities to facilitate science learning in and out of the classroom; perhaps a maker space for teacher ideas and activities. While this may not immediately create a sense of collectivity as with the ILETES teachers, knowing that a safe and regular space exists for them to share with others and further their practice could go a long way in strengthening a collective capacity for dynamic, innovative, and creative-science teaching and learning.

References

Adams, J. (2006). Using the museum as a resource for learning. In K. Tobin (Ed.), *Teaching and learning science: An encyclopedia*. Greenwood publishers.

Adams, J. D., & Gupta, P. (2017). Informal science institutions and learning to teach: An examination of identity, agency, and affordances. *Journal of Research in Science Teaching, 54*(1), 121–138.

Adams, J. D., Miele, E., & Powell, W. (2016). City-as-Lab approach for urban STEM teacher learning and teaching. In L. Avraamidou & W.-M. Roth (Eds.), *Intersections of formal and informal science*. New York: Routledge.

Avraamidou, L. (2014). Studying science teacher identity: Current insights and future research directions. *Studies in Science Education, 50*(2), 145–179.

Charmaz, K. (2005). Grounded theory in the 21st century: Applications for advancing social justice studies. In N. Denzin & Y. Lincoln (Eds.), *The Sage handbook of qualitative research* (3rd ed.). Thousand Oaks, CA: Sage Publications.

Colley, H., Hodkinson, P., & Malcolm, J. (2002). Non-formal learning: Mapping the conceptual terrain, a consultation report.

Gupta, P., & Adams, J. (2010). Museum-University partnerships for pre-service education. In B. Fraser, K. Tobin, & C. McRobbie (Eds.), *Second International handbook of science education*. Kluwer Academic Publishers.

Kenner, C. (n.d.). https://www.llewellyn.com/journal/article/1951

Kohli, R. (2014). Unpacking internalized racism: Teachers of color striving for racially just classrooms. *Race Ethnicity and Education, 17*(3), 367–387.

Martinez, A. Y. (2014). A plea for critical race theory counterstory: Stock story versus counterstory dialogues concerning Alejandra's "fit" in the academy. *Composition Studies*, 33–55.

Mocker, D. W., & Spear, G. E. (1982). Lifelong learning: Formal, nonformal, informal, and self-directed. Information Series No. 241.

National Research Council. (2010). *Surrounded by science: Learning science in informal environments*. National Academies Press.

National Research Council. (2009). *Learning science in informal environments: People, places, and pursuits*. National Academies Press.

Schmidt, B. (2008). The many voices of Caribbean culture in New York City. In H. Henke & K.-H. Magister (Eds.), *Constructing vernacular culture in the trans-Caribbean* (pp. 23–42). New York: Lexington Books.

Solórzano, D. G., & Yosso, T. J. (2002). A critical race counterstory of race, racism, and affirmative action. *Equity & Excellence in Education, 35*(2), 155–168.

· 8 ·

YOUTH PRACTICES AS EXPANSIVISING RESOURCES: DWL ON WHATSAPP

Classroom teachers are often challenged with the use of everyday digital technology in the classroom. For some teachers, smartphones present a distraction, and some teachers or schools even ban smartphone use during instructional time. However, as our cultures become more digitally entangled, where our physical and digital interactions are, it becomes increasingly difficult to separate these two (or more) modes of communication in formal learning environments. As Jahnke's notion of CrossActionSpaces describes, offline and online worlds merge in ways that expand and intersect both classroom and out-of-classroom boundaries. For science teachers in urban schools, this potentially creates expanded opportunities for science teaching and learning that afford students access to voluntary out-of-classroom resources to support their interests in science and encourage their success in institutional measures of science achievement.

Furthermore, as learners choose whether and how to engage in these spaces, their agency in learning expands, something that is often not attributed to racialized urban science learners. In this chapter, I will describe an emergent practice in an urban science teacher's classroom, with the notion of the classroom transcending the temporal and physical spaces that often define it. Specifically, I discuss how the smartphone, in conjunction with a group communication application, extended a teacher's vision of mutual trust and collective success into a sociomaterial space of communication and collaboration beyond his classroom. I employ a hermeneutic approach (Tobin & Richie, 2012), where I analyze the discourse generated in the social media space to make sense of the unfolding social practices in this sociomaterial space. Through this writing praxis, I use my theoretical and experiential lenses to describe and make sense of this teacher's evolving practice and corresponding identity formation vis-à-vis the emergence of this sociomaterial space as both reflection of his identity as a science teacher, his vision of classroom culture and science learning and the salience of this for equitable and meaningful science teaching and learning. Additionally, my stance as a critical researcher compels me to reflect on how this teacher's practice produces spaces of success in science for his Black and Latinx students, who are often left on the margins in terms of access to rich and meaningful science learning experiences (i.e., Adams & Gupta, 2012).

The Collaborative: A Collective Space for Learning About Teacher Identities and Learning to Teach

I have known Lee for almost a decade, first as a teacher education student, then as a near-peer mentor on a grant project and now as a research participant/teacher mentee in my research group. I led the research group entitled Collaborative Teacher Inquiry around Informal Science Learning and Science Teaching in Urban Classrooms (hereafter the Collaborative) that was structured around a National Science Foundation project (of which I was the Principal Investigator), Informal Learning Environments and Teacher Education for STEM (ILETES) to learn about the relationship between learning to teach and teacher identities. Specifically, this project examined how informal science teacher learning experiences influence new teachers' identities and teaching enactments. Teachers invited to participate in the Collaborative were within their first five years of teaching, as learning about their developing identities was important. Furthermore, they all took teacher-certification courses during their undergraduate education that focused on informal science education, with the goal of incorporating informal science resources (museums, zoos, parks, etc.) and approaches in their teaching.

The Collaborative included seven STEM teachers who met on a bi-monthly basis for three years, during which they shared and discussed their developing practices of teaching science. The primary means of data collection during these meetings was cogenerative dialogues (hereafter cogens), structured discourses designed to identify teaching and learning issues and generate new meanings, understandings and solutions or actions around shared experiences (Martin & Scantlebury, 2009). During these meetings, the participating teachers shared their reflections on adapting informal science learning practices into their classrooms, along with corresponding successes and challenges. They also shared artifacts of their teaching that ranged from lesson plans to student work to digital artifacts (photos, videos of their classroom, etc.). Discussions often extended to include the sociocultural factors of teaching in a large, diverse urban district, which is inseparable from the acts of science teaching.

Additionally, most of the teachers in the Collaborative identified as Afro-Caribbean, African American or Latinx. This often showed up in discussions about being a teacher of color in relation to teaching students of color. In the Collaborative, I was both a researcher and professor, as most participants took credit-bearing courses with me during their pre-service teacher education. Moreover, I also taught high school science in the same urban district and then in a museum, where participants in the Collaborative also took courses and professional development.

Therefore, as an Afro-Caribbean woman and a multi-space educator, I was able to engage in what Rita Kohli (2014) terms "reciprocal vulnerability;" as an act of mutual trust, I shared my experiences of science teaching in formal and informal settings as well as my own notions about and experiences with equity, especially for Afro-Diasporic and Latinx students in STEM in urban schools.

Through participation in the Collaborative, the teachers developed collective understandings of informal science learning in relation to the formal classroom and imagined, enacted, and adapted informal science learning in their classrooms to both resonate with their identities as teachers and their visions of student success in their science classrooms. The data I collected included video and audio recordings of the cogens, teaching artifacts, such as lesson plans, photographs, and student work that the teachers brought into the cogens, social media interactions with teachers and field notes of field trips that I facilitated with the participating teachers. Lee emerged as a reflection for this chapter because of his novel use of space in his teaching, both the physical space in the classroom and in the virtual space where the classroom, science learning and youth-centered ways of being and interacting all intersected. Lee consented to my writing about these interactions for this chapter. Here, I will describe the ephemeral yet impactful digital space he and his students co-created for students' success on a compulsory standardized exam. Using the following questions, I reflect on the "innovative" and "unexpected" application of social media in a high school Earth science classroom:

1. In what ways did the use of a social media application resonate with the teacher's identity and vision of science teaching and learning?
2. What were the affordances of a social media application that contributed to mutual trust, collaboration, and student success?
3. In what ways did student and teacher roles and agency emerge and change in the application?

With sociomateriality conveying "an understanding of learning that is situated and embedded within an activity, context and culture and bounded to artifacts making such activity possible (Cerratto Pargman, Knutsson, & Karlström, 2015), this chapter will describe and reflect on the use of the social media platform WhatsApp as situated within the broader context of Lee's unique classroom pedagogy that centers youth ways of being in the world, mutual trust, and self-directed learning, all towards the goal of students' individual and collective success on a compulsory standardized exam.

Identity and Agency and Learning to Teach

Teaching and learning are, in essence, two sides of the same coin. A teacher is always a learner in that they are deepening their knowledge of the subject matter and learning about thierself as a teacher as they interacts with their students in different learning spaces. This also includes the constant re/creation of learning spaces to adapt to thier goals as a teacher and the identities of his students as diverse learners. Nasir, Snyder, Shah, & Ross (2013) describe learning, identity, and goals as relational and evolving together in social practice. The outcome of this is agency, where the learning is defined by how learners access and appropriate resources at hand to meet learning (and teaching) goals. Agency in teaching is the "belief that the self is capable of making the right instructional decisions, knows how to acquire and use resources to teach, and has confidence about constructing and maintaining a safe and effective learning environment" (Adams & Gupta, 2017). However, the agency is not a fixed end-point but rather a constantly evolving entity as a teacher/learner engages in different social practices and spaces. For science teachers, materials are central to creating effective learning environments; having access to laboratory spaces, equipment, physical specimens, and other visual and tactile objects is critical to teaching and learning in the STEM disciplines. However, for science teachers teaching in spaces that are resource-challenged, the agency also means becoming what I am calling a *critical agentic bricoleur*, the ongoing re/creation of resources at hand into new science teaching and learning engagements with special attention to attenuating the challenges faced by students marginalized from science. This newness also includes the appropriation of cultures, such as youths' language and ways of being in social spaces, which are usually relegated to the outside of the classroom into teaching and learning towards student engagement and success. This notion of being a critical agentic bricoleur speaks to the sociomaterial entanglements that constitute STEM teaching and learning—the intersections of physical and digital resources and spaces, bodies, languages, and cultures in the conceptual science classroom.

In this chapter, I illustrate how Lee, as a critical agentic bricoleur, used WhatsApp, a digital app that youth use as a means of communication in their daily lives, as a resource for collaborative science learning; specifically, how he used this platform to confirm group solidarity around success on a high-stakes exam. In this CrossActionSpace that constitutes his physical classroom and the digital space of WhatsApp, Lee created and recreated the collaborative, student-centered, multi-modal learning environment that characterizes his approach to teaching. Also central to Lee's pedagogy is building a solid culture of trust in his classroom. He has a strong belief that if students feel safe and acknowledged, then they will be more open to taking learning risks in the classroom. This risk-taking leads to more positive and successful interactions with science learning. Furthermore, with trust being foundational to his way of being in the classroom, I will show how the emergent practice of WhatsApp as a science teaching and learning resource contributed to the extension of a positive learning culture of mutual trust, collaboration, and academic success in this CrossActionSpace.

Awe and Wonder in the Classroom

Lee is an Earth science teacher in an urban high school. This school is located in a lower socioeconomic community, and as is often the case with such schools, the predominant demographics are historically racialized and minoritized students who have experienced years of subpar schooling and have, as a result, largely disengaged. This is evident in the state test results where Lee reported that there had been consistently low passing rates; in the year prior to Lee's appointment, it was less than 10 percent. These tests are required for graduation, so students either repeat the class until they pass or get pushed out of school. Being a graduate of the city's public school system and a minoritized person himself, Lee empathizes with his students. He maintains his commitment to teaching in New York City despite administrative pressures. In one Collective meeting, he described the reduced pressures he would have teaching in the suburbs, including not having to worry about low passing rates,

> "But then, as I was reflecting on it, that's not why I became a teacher. That's the easy route; you know, if I do that, I'm going to become a sell-out. No. I need the challenge. I need to help the individuals that remind me of my friends and my situation growing up. So, I stuck to the school, and I think I'm going to just ride it out for a few more years and just stay there."

However, more than riding it out, Lee actively imagined and created novel ways to keep his students engaged in science learning, including the use of communication technology as a means of sustaining the social bonds, both to each other and to the subject, which developed in the classroom. Lee has a natural way of integrating students' ways of being and communicating in his teaching, and this creates spaces for success in science for students for whom this success is often elusive. To an outside observer, Lee's classroom may seem noisy and even a little chaotic. However, a close look would reveal a careful structure where students are engaged in science learning but allowed to be "teenagers"—highly vocal, physically active and exchanging jokes with each other and Lee. As an urban denizen myself, Lee's classroom reminds me of the ordered chaos of the city's streets where people appear to be random and engaged in diverse activities but there is a collective direction—everyone has a place to go and a path to get there. In later sections of this chapter, I will share some vignettes to illustrate Lee's teaching and learning culture.

During the Collaborative meetings, Lee often lamented about the numbers and how his success as a teacher was judged by passing rates, no matter how innovative his teaching and engaged his students. As a new teacher, Lee saw this constraint as an opportunity for expansive teaching, learning and agency in his classroom. Because the Collaborative was structured around informal science learning, Lee queried in an early group meeting,

> "Why can't we just put the informal inside the classroom? Why does it have to be outside the classroom? When you think of what it is, we are learning, we're relaxed, we're critically thinking, why can't we just do that in the classroom?"

Like other members of the Collaborative, Lee's teacher education at the public liberal arts college included informal science learning through courses in collaboration with a natural history museum and various summer activities structured through the pre-service teacher education program. He and others in the group reflected on these as being meaningful learning experiences for them as teachers and strived to recreate similar learning experiences with their students. However, most of this emphasized reproducing fieldtrip-type experiences. Lee challenged the group to recreate meanings of informal science learning for the formal, urban science classroom. Lee wanted to have "awe and wonder" *in* his classroom and not just on field trips.

Being an "Informal" Teacher

Lee identifies himself as an "informal science" teacher; he mentions this whenever he describes his pedagogy or simply states the kind of teacher he is. This stands in contradiction to his emphasis on the standards and assessments that both judge him as a teacher and his students as learners. In the urban district where he works, the teachers are subjected to stringent rating systems based on the Danielson Framework (Danielson, 2011). Although Danielson herself has denounced this subjectification of teachers (Danielson, 2016), this framework persists, and some teachers view it as a mechanism to undermine their sense of agency in the classroom. Lee, for one, has always mentioned this framework as both a goal and a constraint to his teaching. During a goal-setting meeting at the beginning of a term, he mentioned having a "highly effective" rating as his goal, along with successful classroom management and good relationships with his students. For him, maintaining a student-centered environment is key to achieving these goals,

> "You know…I have a vision of what they should be doing and what I want them to do, but I want to present it in a way that is informal. It's creating that environment where they can drive and succeed, but at the same time, they're choosing their path to get the result that…I present in the beginning."

He related this to norms of visiting informal science institutions,

> "When you first asked us how we talk about informal [science], it's a museum and an exhibit and a zoo. You know what is, the first thing they had you, they hand you a tri-fold paper with the lists of all the exhibits, a map. And you are free to wander and learn in that environment and go anywhere. And they have a plan. The plan is [that the visitor] leave this area, the zoo or this exhibit learning A, B, and C. But you can take any path to achieve that. Yeah, I really like that."

Since his initial years of teaching, he has used this approach to shape how he structures teaching and learning in his classroom. He developed multiple modes and pathways for his students to achieve the learning outcomes that he planned. He positions himself as a coach and mentor, "I want to stay on the side. I don't want to be a heavy-handed teacher," Lee's way of saying that he'd rather be the "guide on the side" than "sage on the stage." Because this kind of structured informal learning is central to Lee's identity as a teacher, he makes use of a variety of resources at hand to shape and enable student-centered and self-directed learning.

His teaching and learning, although structured by standardized tests, afforded flexible learning spaces for his students and multiple ways of engaging in science. He had a penchant for leveraging the ways that adolescents engage with their worlds in the classroom. For example, Lee painted old lab tables with chalkboard paint and rendered writing on the desks as a legitimate learning activity in his classroom. In between the random doodles and graffiti tags are scribblings and sketches of Earth science concepts. Similarly, he uses "tagging" as a review activity; students use dry-erase markers to draw their understandings on the windows and engage in gallery walks to learn and review each other's drawings. A number of videos that Lee posted about his classroom on Facebook show students singing, and one in particular shows a student going back and forth between playing on a synthesizer and working with his group. Amidst these activities, students also demonstrate engagement in science by breaking verse to ask a salient question or moving to another table to help out classmates. These actions in the classroom convey a strong sense of mutual trust, which is important in generating success in any learning environment, physical or digital.

Centering Trust

The activities that Lee enacts in his classroom serve not only as a means for student learning but also as a way to build solidarity and trust in the classroom, which is critical to student success. Developing trust in the classroom is a critical yet underemphasized aspect of learning to teach. The relationships that teachers build with students are just as important as the quality of the subject taught (Woodson & Id-Dean, 2016). Lee emphasizes relationship-building with his students,

"Every kid that comes into my class, I have a dialogue with them. I have a relationship with them…you know, I meet them at the door, and I greet them. And at that very instant whatever, whatever they choose to respond with me, I know we're going to have a smooth day. I know the student is having an issue with something, and I have to find out. The faster I find out what the issue is, if they're hungry or … if they didn't sleep or if they have to work,…then I have to fix that issue because once I get rid of that distraction and fix that issue, then the rest of my class, rules, management, and instruction become smooth, and that student buys in. Because they know I care. They know they can talk to me."

Catherine Ennis and Terri McCauley note, "Trusting environments are best created in classrooms in which students and teachers can work co-operatively over an extended time period to construct trusting relationships" (2002, p. 151). In the classroom, trust extends beyond socioemotional safety and towards a belief in all students' abilities, which is that all students can learn. Id-Deen and Woodson (2016) found that the absence of trust negatively affects learning. Creating a trusting environment includes rigorous learning opportunities, creative teaching and learning strategies and allowing for on-task dialogue. Trust also includes allowing students to take ownership of their own learning; Lee describes his alternative to station teaching,

"I do a café-type model where I construct an alternative: four different paths for the students to go through the stations because determining what they answer to the do now, their path is constructed."

He is also transparent in his grouping as he lets students know why they are grouped the way they are grouped and follow the path that they follow through the stations. Likewise, he structures the cafés so that once higher performers finish, they move to a path where they help the lower-performing students,

[If] my lower-performing students display an understanding or mastery of the concept we are reviewing, or we are going to do or even think forward about that concept, then I know I am going to have a successful lesson. Because then I can assume most of my students understand it."

In Lee's classroom, students play a central role in helping their peers, and this is something that he frequently reinforces by reminding them that in his class, "We all help each other out." In a video Lee posted on Facebook, he noted, including a jab to the Danielson Framework, "[This is what] highly effective teaching looks like; I didn't say one full sentence that whole class.

Students lead work and explain concepts; help each other. Everyone here did amazing," he exclaimed.

Interaction Rituals and Learning

The video that Lee posted demonstrated the entrainment and corresponding solidarity in his classroom. The students were freely moving around to different tables, and while there was "off-task" joking, it was clear that the students remained on-task and that the joking contributed to the playful atmosphere that kept them engaged and focused. Lee allows his students the freedom to engage in learning in ways that resonate with who they are as learners and how they wish to engage in the tasks at hand. These activities serve to generate consistent positive emotional energy, an emotional state of feeling empowered, confident, and agentic (Seiler & Elmesky, 2007; Collins, 2004), in the classroom, which in turn contributes to productive science learning.

Kenneth Tobin and colleagues extended Randall Collins' (2004) notion of Interaction Ritual Chains (IR) to science teaching and learning. Through research and enactment of successful science teaching in urban schools, they established that the generation of positive emotional energy through social interactions is important to successful classroom teaching and learning, "If a teacher can be involved with the creation of chains of successful encounters, then rituals can occur with which participants can associate positive emotions, just by thinking back on the activity" (Tobin, 2004, p. 15). This perspective also articulates the importance of mutual respect, collective effervescence, affiliation, and solidarity in student success. Stacy Olitsky (2007) describes,

> "Emotional Energy is generated through successful interaction rituals that are characterized by bodily co-presence, mutual focus, common mood, boundaries to outsiders, an "entrainment," or coordination of body movements and speech, shared experience between participants on both an emotional and cognitive level and solidarity with others present." (p. 8)

While much of this work was done in a face-to-face context, reflecting on the emotional component of interactions in CrossActionSpaces will be important in describing learning in the digital age. Lee's identity as an informal science teacher who values student-centered, multi-modal learning also shaped how he used WhatsApp in his teaching. He incorporates this resource for a very specific goal—student success on the standardized tests—yet maintains the culture of trust and peer learning that characterizes his classroom. As I will demonstrate, the evidence of successful interactions and positive emotional energy was very specific to this communicative space but transcends the boundaries of the digital space into a CrossActionSpace that incorporates the physical and temporal classroom and the ephemeral conceptual and digital spaces that are integral to the lived experiences of Lee and his students.

Extending Interaction Rituals to CrossActionSpaces

> So, WhatsApp is a chat app, right? You can create a chat room using the app and download it. I've been using it because I'm letting the students access me outside of school. So, when I'm home if they have a question about homework or a concept we did in the morning or during lessons, they can ask me. There are about thirty students in the room. I gave my number to all of the students, but not everyone downloaded the app to message me. And I don't force them to. I review the concepts and WhatsApp allows me to do a number of things. It allows them to have access to me at any time of night. It allows me to send them videos from my house while I am going over the concept with them, while I am going over the concepts, I direct the video. In my room [at his house], I put whiteboard paint on it so I can draw things while I am talking to them. Or write, while I am talking to them and explaining. Then they have dialogues with each other, sharing [re]sources. That's how I've been using it in my class because that is the only time they get this instruction [and] they don't have good study habits, so this gets them exposed more to me and my content when they are outside my classroom.

For Lee, the classroom does not end at the last bell of the day and at the boundary of the physical plant. Rather, the classroom goes beyond these temporal and spatial borders with WhatsApp. Lee recognizes the challenges that his students have in maintaining good study habits at home and views his role as ensuring that they have a space to develop and hone productive study habits with his guidance. He views WhatsApp as a means for his students to access additional help in science outside of school time and as a way of deepening his personal connection with his students. Although research on the use of WhatsApp in teaching and learning contexts is emergent, researchers have learned that WhatsApp provides a useful platform for connecting students during out-of-school time. For example, Nirgude and Naik (2017) described it as an effective tool for sharing information, engaging in discussions, assessing learning, and offering feedback. In another study, Sayan (2016) found that WhatsApp contributed to student achievement on exams with both instant knowledge and the use of the media as motivating factors. Bouhnik and

Deshen (2014) interviewed secondary teachers about their use of WhatsApp in the classroom and learned that the technology served four purposes: communicating with students, nurturing social relations, creating dialogue, and encouraging sharing amongst students and as a learning platform. These findings indicate that using social media platforms, such as WhatsApp, in education has the promise of extending the learning environment of the classroom beyond the school, both in time and space. While these studies have examined the features that make WhatsApp conducive to teaching and learning, they did not examine the dialogues that contributed to learning, including the social relationships that are critical in maintaining a safe, trusting, and effective learning environment within the WhatsApp interactions.

In the following sections, I present several dialogues from Lee's WhatsApp interactions with his students and describe the ways that these dialogues contributed to the solidarity and positive emotional energy in the group. I highlight the features of the dialogues, as they emerged in the app, which extend the solidarity that characterizes Lee's physical classroom and contributes to successful science learning amongst his students.

Collaborative Learning in WhatsApp: A Low Stakes Environment for a High Stakes Review

In early June, Lee established a WhatsApp group entitled State Exam Prep[1]; he included me in the group because of our relationship in the Collaborative. It began with a series of him adding students' numbers to the group interspersed with emoticons and greetings like "Yoooo" and "Whassup" from students acknowledging their admittance into the group (on my end, I all of a sudden received a tidal wave of additions and greetings notifications via my WhatsApp, it was not until Lee texted me separately, I realized that I was a part of the State Exam Prep group). Greetings perform a critical role in the establishment and maintenance of social relationships (Li, 2010), so the greetings in WhatsApp were the first step in marking this virtual space as an extension of the relationships developed in the classroom. Some of the greetings were general to the whole group, whereas others were specific either to the teacher, "Whassup Lee," or student-to-student, "hey girl." Although the participation in the group was voluntary and the review happened in the evening, there were an impressive number of students involved, as indicated by unique colors and phone numbers in the app. In the representative dialogues,

I used the last four digits of students' phone numbers to signify the different voices.

Each evening was for the review of a different Earth science topic. Lee established the ritual of sending out an IM letting students know that the first video would be posted at 7 pm so that they could anticipate the session. (In one of the sessions, a couple of students entered the space early, around 6:45 and were quickly reminded by Lee and others of the first post-time ritual). The video was usually an explanatory video of a topic covered in class during the prior year and would be on the State Exams. The culminating post was often a picture of a review sheet with sample questions from the State Exams. In exchange for donuts, Lee asked them to submit the questions directly to him by 6 am the following morning. During his review, Lee also posted questions for students to answer and discuss.

The communicative style in WhatsApp is very informal. The text includes slang, typos, phonetic spellings, creative uses of grammar, and hegemonic English. Similar to in the classroom, Lee and his students seamlessly code-switch between hegemonic English and vernacular English (and there is a digital English that seems to be more related to spoken vernacular English; in addition to emojis, it includes abbreviations and phonetic-like spellings that appear to be used to convey speech emphases); these fluid changes in speech structures allow for the attenuation of borders between the classroom and the surrounding community and establish a sense of community within the classroom. Being able to communicate freely in students' (and teachers') home and school languages also serves to solidify the trust and social bonds in this virtual space. The use of emojis is also a key feature in WhatsApp communication and serves as sources of entrainment amongst Lee and his students.

In the following example, Lee presents the hydro[logic] cycle through a series of short, informal videos. The post is followed by sample State Exam questions, followed by "enjoy" and a smiley emoji. This signals to the students that this is a low-stakes space to help them to pass the exams:

| 3603 | Okay So The Water From Underneath The Ground Is Causing The Run Off. Why Does It Start Specifically In That Area? |

Lee	-[Answer runoff video]
	-[Diagram hydrocycle]
	-[Scan of test questions]
	-And here are the regents questions enjoy [smiling emoji]

Students immediately follow with questions about key terms and queries about where to submit their answers. Lee jokingly responds which is followed by two laughing students:

9230	-What's percolation?!?
	-Do we post our answer's in the group?!! Rn
Lee	Send me the answers in a private msg so M won't copy them [two laughing emojis]
9230	Dwl* ohk
3603	-[Three laughing emojis]
	-What does permeable mean?

*dead wid laugh

Another student posts screenshots of definitions of the word in question; Lee praises the group, "I love it when you help each other out." More joking happens and the entrainment of the laughing emojis ensues:

5576	-[screenshot definition of permeable]
	-[screenshot definition and diagram of percolation]
9230	Thank uh
Lee	[2:50 min video of him explaining permeability using the whiteboard]
3603	Thank you
Lee	-[Two laughing emojis]
	Good job [5576] [three laughing emojis]
5707	I couldn't respond to anything right now because I'm trying to watch your vids [sad emoji]
Lee	It's ok [ok symbol emoji] just focus

0357	Did I get the answers right??
8177	So when the soil is saturated run off occurs??
5707	Yes.
Lee	[Smiling emoji] I love it when I see you all help each other
6656	Can we help each other on the [State Exams]?
0530	[three laughing emojis]
Lee	[five emojis]
3603	I'm Dead [laughing emoji]
8177	[two laughing emojis]
3507	[two different laughing emojis]
9633	[one laughing emoji]
7165	[three laughing emojis]
5707	Is everybody going crazy here?!?
Lee	N my buddy we just bonding and working on [State Exams] and happy we Getting it done
4083	[Sideways laughing emoji]

As testing is usually a source of stress and anxiety for students this series of laughter serves as both a release and the generation of positive emotional energy. The emojis signal a collective effervescence indicative of successful interactions (Ritchie, Tobin, Hudson, Roth, & Mergard, 2011). Lee even articulates this feeling when he writes, "we just bonding and working...and happy we getting it done."

In the exchange below, the students review their answers to the questions. There are consolations and discussions about reasons for the wrong answers and collective praise for the right answers. These are also factors that demonstrate group solidarity, as the students are able to console each other's failures and celebrate successes. Lee "contributed" a reward to this exchange, but the affective factors, including positive emojis (smiling and clapping hands) of this exchange, were student-to-student. Furthermore, the students feel a level of trust and safety in being able to share their "failures" without fear of being ostracized by Lee or other classmates. In this learning space, the students' agency changes as they take more ownership of the direction of the learning exchanges and Lee's role is configured differently from that of the teacher to that of a facilitator, which resonates with the way that he desires to be as a teacher. In the segment below, one student lamented about getting one wrong that was consoled with "hushies" from another student showing that were was a certain degree of care for her classmate:

5791	Well i got number 2 wrong the answer is (3) because permeable it the ablity to allow the soil soak in water
Lee	So impermeable means to not soak up water...
5791	Yes
9230	I got all right [clapping hands and smiling emoji]
8177	[Clapping emoji]
9230	Yesh lexiie
Lee	When everyone does the day's work and vids you get a reward
6656	Everyone*
8177	I got one wrong [sad emoji]
9230	Hushies.. You'll get it next time
7620	You guys are so motivational
9230	[Two winking emojis] Thanks [7620]
0764	Anyone pree I did the homework
9230	We got homework [rolling eyes emoji]
3603	Got job [0764] [shout out emoji]
0764	Thank you [grateful emoji]

Another exchange demonstrated a breach of trust when one student wanted Lee to confirm his answer rather than his classmates:

3174	Ya
3043	I want Lee to tell me that I'm right just to be sure [thinking emoji]
Lee	Ur rite

3174	[three angry and three steaming emojis]
3043	[grateful smile emoji]
Lee	Red shift only
3043	-Got you -Copy
Lee	-We are here to help each other [3043] - Listen to them
3174	[two angel emojis]

Although several of his classmates said, "red shift," which was the correct answer, he still wanted confirmation from Lee. This was met by angry emojis from one of the students who posted the correct answer. Lee reminded them of the class culture, which was followed up with angel emojis from the same student who displayed annoyance at being ignored.

In this low-stakes environment, students are also more motivated to work out the science problems themselves rather than getting the answers from others. Lee creates a positive culture of learning and achievement in his practice, which extends to his students wanting to learn science and wanting to do well. In the exchange below, one student in particular got upset when another did not follow the instructions of submitting the answers directly to Lee,

Lee	All answers are due by 8:15 today's a easy day [heart and globe emoji] if you can't submit by then -I send u a different vid
9230	-I feel soo smart I got everyone right [three tears-of-joy emojis] -Im soooo happy -Everything*
8177	Nice
5707	39 is 3) 12.5 hours!
9230	Private
5791	So you aint see the "send me the answers in a private message"
5707	-46 is 1) 11m! -47 is 4) perigee and a Full-Moon phase!
5791	[5707] SEND HIM A PRIVATE MESSAGE

| 5707 | How am I gonna do that |
| 5791 | -Your not giving everyone a chance to figure it out on they own r u blinddd?!
 -Text him on whats apps duh
 -Just like everyoneee else |

The conversation continued with 5707 apologizing, another student chiming in, "It's Okay [5707] I Made That Mistake Too," before there was a collective instructional on how to send Lee a private message. With the negative emotional energy building as the two students continued the heated exchange, Lee interjected with a humorous diffusing video of himself, which elicited laughter from the students:

| 9230 | [five laughing with tears emojis] am dead |
| 8177 | [Skull emoji] |

A student even made a screenshot of a funny face Lee made during his video and shared it with the group. This was followed by more laughter and "dead," signaling a redirection of the energy in the group. According to Olitsky (2007), "During successful interaction rituals, the symbols that are both created and exchanged become invested with positive emotional energy and can be used later to generate successful interaction rituals with others who find these symbols similarly charged" (p. 36). This screenshot became a symbol that reemerged at random times during different sessions, eliciting laughing emojis and "dwl" from the group each time. It was a recurring resource for eliciting collective effervescence and positive emotional energy and reinforced the solidarity in the group. These small symbols produced successful interactions and reinforced the overarching symbol of achievement in science and success on the State Exams. The symbol of success brought the students to the WhatsApp reviews each evening, and the symbols that emerged during the chat encouraged them to stay.

From the initial interactions on WhatsApp to the final weeks of the review, which were about three weeks, the emotional energy in the chats largely remained positive and supportive. Lee and the students managed to quickly diffuse disagreements that arose. This review and this space were important for the students; they made a collective effort to keep the space focused and positive. The irony to me, as an outside observer to this chat, was that many of the students were not fully aware of who all of the others were in the chat until a few sessions in (in WhatsApp, some of the students used their real names whereas many used nicknames). Lee explained to me that several sections of the same class were in the chat, so not all of the students knew each other. During a session about a week into the reviews, there was a series of "who's who" exchanges to identify those with unknown nicknames and students from different sections of the same class. This is evidence of the trust that the students have in both Lee and each other, and they could enter this space of "strangers" and fully engage in the learning processes without fear of being humiliated for being "wrong" or "failing".

Configuring a Culture of Collaboration with WhatsApp

The WhatsApp communication allowed for multi-modal engagement in science learning that included learning the science content and processes and solidifying social bonds around science. There was a strong emphasis on solidarity building through joking, emojis, collective encouragement and admonishments when the social norms of the class were broken. Both Lee and the students created and sustained a strong culture of positive emotional energy around science learning. One student even joked about hating school to see the group's reactions, knowing that hating school was counter to the learning culture of Lee's classroom.

3174	I really hate school [four angry and three steaming emojis]
9230	Dwl why
5707	That means you ain't never gonna be smart!
3174	-Im jk i just wanted to see yall reaction [two cool emojis] -[two big smile emojis]
5707	If you hate school, you'll never be intelligent in the future! School is important in life!
3174	Yea
Lee	Yes you see in my class I teach science but always about life
5707	Are we ready for your vids, Lee

6656	[two laughing emojis]
3174	-Its 7 no -Now*
5707	I'm ready! [three star emojis]

"Online encounters…are…constitutive of an ensemble of encounters that comprise our various relationships in and through the real and the virtual." As stated in their research on WhatsApp usage, O'Hara (2014) and colleagues describe a CrossActionSpace; the social relations therein are reflective of the relationships that we develop in our various and intersecting life spaces. They describe it as a digital age dwelling: "human affairs entail a movement through and between sites of engagement, where trajectories of individuals intersect and create a texture of joint being together, a felt-life of sociality" (p. 1133). In CrossActionSpaces, the divide between the virtual and physical is dissolved, and learning cultures are transcendent. The WhatsApp space decentralizes power in teaching and learning and affords students agency in how they direct the learning space. There is a distributive sense of agency where both participating students and Lee share responsibility for the collective learning, the generation and maintenance of positive emotional energy and a collective solidarity towards success on the defining exam.

In WhatsApp, as an extension of Lee's classroom, science learning is a part of the social activity, one in which both the teacher and students look forward to engaging because of the positive emotions it generates. It is the reiteration between the interaction rituals (the consistent time of the group meetings and starting with videos and exam questions, dialogue exchanges that extend the informal nature of communicating in the classroom into the digital space and collective emoji use to signify positive and some negative exchanges within the dialogues) and trust that make the WhatsApp space a successful learning space as an extension of Lee's classroom. Through these modes of interaction, the students configure this as a social space, and the symbols contextualize the students' connection with Lee, each other, and the science content and exam. The symbols created in the classroom transfer into the WhatsApp space and contribute to the success of this space in achieving the goals of student science learning.

Both Lee and his students' identities are shaped by the learning culture in the CrossActionSpace that intersects their classroom, community and the digital spaces in which they live. Through the shaping of the WhatsApp space to afford youth agency in shaping the direction of the learning interactions, Lee creates a resource for the students to develop positive identities around school learning, science and academic success, identities that are often not available to urban students of color (Nasir et al., 2013). As a teacher, Lee's identity as an "informal teacher" compels him to develop and maintain learning spaces where his students choose their own path to learning and drive themselves and each other towards science/academic achievement. In the case of both Lee and his students, the agency developed in this CrossActionSpace is transcendent and creates the conditions for all participants to develop identities that resonate with imaginations and realizations of academic advancement.

Lee's teaching philosophy drove how he configured this digital space for his classroom. He created the conditions for trust and solidarity in his classroom that extended to his use of WhatsApp; and the affordances of WhatsApp allowed for an extension of Lee's culture of collaboration. As a social media application, Whatsapp has key affordances that make it a salient resource for transcending a physical classroom space. It is a free app. With the widespread use of smartphones in everyday life, this app is available to everyone and affords rapid and asynchronous or synchronous communication between users. Lee scheduled a time, and many students were "present" to participate in the review. However, the dialogues and videos remain in the space, allowing latecomers or those who could not do the review to still benefit from the learning discussions that occurred. This application allows for multimedia communication—photos, videos, voice messages—which is salient for education. In this space, Lee employs a cogenerative and emergent version of "just-in-time" teaching (i.e., Novak, 2011) where there is a simultaneous production, assessment and reproduction of knowledge. Lee introduced materials in this space, of which some students had some prior knowledge, and students' responses immediately alerted Lee to their levels of understanding and gaps in knowledge that needed to be addressed. Because Whatsapp is multimodal, Lee was immediately able to produce resources to meet students' learning needs. This is not always possible in the classroom, with set lesson plans and curricula and rigid time structures that often impede flexibility in teaching.

The flexibility of this CrossActionSpace also afforded the students agency in determining the direction of the sessions, spending more or less time on a given topic depending on their satisfaction with their levels of understanding, which was uninhibitively expressed with emojis. Although Lee initiated these spaces to improve his students' chances of success on the state exams, the space has always and immediately become a cogenerated and collaborative learning space. Lee and his students were equally invested in community building and science knowledge production. In an urban school that was often plagued with low test scores and disinterested students, Lee and his students created this space where science learning and success were a critical and collective goal.

In the 21st century, we have access to numerous digital tools for communication and collaboration. Not only are these tools being rapidly developed, but they have also augmented our ways of communicating and relating. While we readily use these tools in our day-to-day interactions, applying these tools to teaching and learning settings would go a long way to blurring the boundaries between schools and community and allowing for more informal interactions to not only contribute to classroom knowledge production but also fostering the emotional bonding that is necessary for creating safe, trusting, and effective learning spaces.

A version of this chapter appears in Pargman, T. C., & Jahnke, I. (2019). Emergent Practices and Material Conditions in Learning and Teaching with Technologies. *Cham, Schweiz: Springer.*

Notes

1 In the United States, different states have different policies regarding standardized exams and secondary school graduation requirements. In the state where this study took place, secondary school students are required to take and pass subject-area state exams, including Earth science, in order to get a high school diploma.

References

Adams, J. D. (2014). Place and identity: Growing up bricoleur. In K. Tobin, & A. A. Shady (Eds.), *Transforming urban education: Collaborating to produce success in science, mathematics, and technology education*. Rotterdam, NL: Sense Publishers.

Adams, J. D., & Gupta, P. (2017). Informal science institutions and learning to teach: An examination of identity, agency, and affordances. *Journal of Research in Science Teaching, 54*(1), 121–138.

Adams, J. D., & Gupta, P. (2013). "I learn more here than I do in school. Honestly, I wouldn't lie about that": Creating a space for agency and identity around science. *The International Journal of Critical Pedagogy, 4*(2).

Bouhnik, D., & Deshen, M. (2014). WhatsApp goes to school: Mobile instant messaging between teachers and students. *Journal of Information Technology Education: Research, 13*(1), 217–231.

Cerratto-Pargman, T., Knutsson, O., Karlström, P. (2015). Materiality of online students' peer-review activities in higher education. In Proc. of CSCL 2015, pp. 308–315.

Collins, R. (2004). *Interaction ritual chains*. Princeton university press.

Danielson, C. (2016). Charlotte Danielson on rethinking teacher evaluation. *Education Week, 35*(28), 20–24.

Danielson, C. (2011). *Enhancing professional practice: A framework for teaching*. ASCD.

Ennis, C. D., & McCauley, M. T. (2002). Creating urban classroom communities worthy of trust. *Journal of Curriculum Studies, 34*, 149–172.

Id-Deen, L., & Woodson, A. N. (2016). "I Know I Can Do Harder Work": Students' perspectives on teacher distrust in an urban mathematics classroom. *Urban Education Research & Policy Annuals, 4*(2).

Jahnke, I. (2015). Digital didactical designs. Teaching and learning in CrossActionSpaces. New York: Routledge.

Kohli, R. (2014). Unpacking internalized racism: Teachers of color striving for racially just classrooms. *Race Ethnicity and Education, 17*(3), 367–387.

Li, W. (2010). The functions and use of greetings/LES FONCTIONS ET L'UTILISATION DES SALUTATIONS. *Canadian Social Science, 6*(4), 56.

Martin, S. N., & Scantlebury, K. (2009). More than a conversation: Using cogenerative dialogues in the professional development of high school chemistry teachers. *Educational Assessment Evaluation and Accountability, 21*(2), 119–136.

Nasir, N., Snyder, C. R., Shah, N., & Ross, K. M. (2013). Racial storylines and implications for learning. *Human Development, 55*(5–6), 285–301. doi: 10.1159/000345318

Nirgude, M., & Naik, A. (2017). WhatsApp application: An effective tool for out-of-class activity. *Journal of Engineering Education Transformations*.

Novak, G. M. (2011). Just‐in‐time teaching. *New Directions for Teaching and Learning, 2011*(128), 63–73.

O'Hara, K. P., Massimi, M., Harper, R., Rubens, S., & Morris, J. (2014, February). Everyday dwelling with WhatsApp. In *Proceedings of the 17th ACM conference on computer supported cooperative work & social computing* (pp. 1131–1143). ACM.

Olitsky, S. (2007). Promoting student engagement in science: Interaction rituals and the pursuit of a community of practice. *Journal of Research in Science Teaching, 44*(1), 33–56.

Ritchie, S. M., Tobin, K., Hudson, P., Roth, W. M., & Mergard, V. (2011). Reproducing successful rituals in bad times: Exploring emotional interactions of a new science teacher. *Science Education, 95*(4), 745–765.

Sayan, H. (2016). Affecting higher students learning activity by using WhatsApp. *European Journal of Research and Reflection in Educational Sciences, 4*(3).

Seiler, G., & Elmesky, R. (2007). The role of communal practices in the generation of capital and emotional energy among urban African American students in science classrooms. *Teachers College Record, 109*(2), 391–419.

Tobin, K., & Ritchie, S. M. (2012). Multi-method, multi-theoretical, multi-level research in the learning sciences. *Asia-Pacific Education Researcher, 21*(1).

Tobin, K. (2007). Collaborating with students to produce success in science. *Journal of Science and Mathematics Education in Southeast Asia, 30*(2), 1.

Studies in Criticality

Series Editor
Shirley R. Steinberg

Counterpoints publishes the most compelling and imaginative books being written in Education and Cultural Studies today. Grounded on the theoretical advances in critical theory, feminism, and postcolonialism in the last two decades of the twentieth century, Counterpoints engages the meaning of these innovations in various forms of educational expression. Committed to the proposition that theoretical literature should be accessible to a variety of audiences, the series insists that its authors avoid esoteric and jargonistic languages that transform educational scholarship into an elite discourse for the initiated. Scholarly work matters only to the degree it affects consciousness and practice at multiple sites. The editorial policy of *Counterpoints* is based on these principles and the ability of scholars to break new ground, to open new conversations, to go where educators have never gone before.

For additional information about this series or for the submission of manuscripts, please contact:

> Shirley R. Steinberg, Series Editor
> msgramsci@gmail.com

To order other books in this series, please contact our Customer Service Department:

> peterlang@presswarehouse.com (within the U.S.)
> orders@peterlang.com (outside the U.S.)

Or browse online by series:

> www.peterlang.com

www.ingramcontent.com/pod-product-compliance
Lightning Source LLC
Chambersburg PA
CBHW061714300426
44115CB00014B/2687